Introduction to
Quantitative Data Analysis
in the Behavioral and
Social Sciences

Introduction to Quantitative Data Analysis in the Behavioral and Social Sciences

Michael J. Albers
East Carolina University

Registered Offices
John Wiley & Sons, Inc., 111 River Street, Hoboken, NJ 07030, USA

Editorial Office
111 River Street, Hoboken, NJ 07030, USA

For details of our global editorial offices, customer services, and more information about Wiley products visit us at www.wiley.com.

Wiley also publishes its books in a variety of electronic formats and by print-on-demand. Some content that appears in standard print versions of this book may not be available in other formats.

Library of Congress Cataloguing-in-Publication Data applied for.
Hardback: 9781119290186

Cover image: Magnilion/gettyimages
Cover design by Wiley

Set in 10/12 pt WarnockPro-Regular by Thomson Digital, Noida, India

10 9 8 7 6 5 4 3 2 1

Contents

Preface

This book strives to be an introduction to quantitative data analysis for students who have little or no previous training either in statistics or in data analysis. It does not attempt to cover all types of data analysis situations, but works to impart the proper mindset in performing a data analysis. Too often the problem with poorly analyzed studies is not the number crunching itself, but a lack of the critical thinking process required to make sense of the statistical results. This book works to provide some of that training.

Statistics is a tool. Knowing how to perform a t-test or an ANOVA is similar to knowing how to use styles and page layout in Word. Just because you know how to use styles does not make you a writer. It will not even make you a good layout person if you do not know when and why to apply those styles. Likewise, statistics is not data analysis. Learning how to use a software package to perform a t-test is relatively easy and quick for a student. But knowing when and why to perform a t-test is a different, and more complex, learning outcome. I had a student, who had taken two graduate-level business statistics courses, remark when she turned in a statistics heavy report in a writing class: "In the stat classes, I only learned enough to get me through the test problems. I have no idea how to analyze this data." She had learned how to crunch numbers, but not how to analyze data. Bluntly, she wasted her time and money in those two classes.

The issue for researchers in the social sciences is not to learn statistics, but learn to analyze data. The goal is not to learn how to use the statistical tests to crunch numbers, but to be able use those tests to interpret the data and draw valid conclusions from it. There is a wide range of statistical tests relevant to data analysis; some that every researcher should be able to perform and some that require the advice/help of a statistical expert. Good quantitative data analysis does not require a comprehensive knowledge of statistics, but, rather, knowing enough to know when it is time to ask for help and what questions to ask.

Every quantitative research study (essentially by definition) collects some type of data that must then be analyzed to help draw the study's conclusions. A great study design is useless unless the data is properly analyzed. But teaching that data analysis to students is a difficult task. What I have found is that most textbooks fall into one of these categories.

- Research method textbooks that explain how to create and execute a study, but typically are very light on how to analyze the data. They are excellent on explaining methods of setting up the study and collecting the data, but not on the methods to analyze it after it has been collected.
- Statistics textbooks that explain how to perform statistical tests. The tests are explained in an acontextual manner and in rigorous statistical terms. Students learn how to perform a test, but, from a research standpoint, the equally important questions of when and why to perform it get short shrift. As do the questions of how to interpret the results and how to connect those results to the research situation.

This book differs from textbooks in these two categories because it focuses on teaching how to analyze data from a study, rather than how to perform a study or how to perform individual statistical tests. Notice that in the first sentence of a previous paragraph I said "data that must then be analyzed to help draw the study's conclusions." The key word in the sentence is help versus give the study's conclusions. The results of statistical tests are not the final conclusions for research data analysis. The researcher must study the test results, apply them to the situational context, and then draw conclusions that make sense (see Figure 1.1 in Chapter 1). To support that process, this book works to place statistical tests within the context of a data analysis problem and provide the background to connect a specific type of data with the appropriate test. The work is placed within long examples and the entire process of data analysis is covered in a contextualized manner. It looks at the data analysis from different viewpoints and using different tests to enable a student to learn how and when to apply different analysis methods.

Two major goals are to teach what questions to ask during all phases of a data analysis and how to judge the relevance of potential questions. It is easy to run statistical tests on all combinations of the data, but most of those tests have no relevance or validity regardless of the actual research question.

This book strives to explain the when, why, and what for, rather than the button pushing how-to. The data analysis chapters of many research textbooks are little more than an explanation of various statistical tests. As a result, students come away thinking the important questions are procedural, such as: "How do I run a chi-squared test?" "What is the best procedure, a Kruskal–Wallis test or a standard ANOVA?" and "Let me tell you about my data, and you can tell me what procedure to run." (Rogers, 2010, p. 8). These are the wrong questions to be asking at the beginning of a data analysis. Rather, students need to think along the lines of "what relationships do I need to understand?" and "what are the important practical issues I need to worry about?" Unfortunately, most data analysis texts get them lost is the trees of individual tests and never explains where they are within a data analysis forest.

Besides knowing when and why to perform a statistical test, there is a need for a researcher to get at the data's deep structure and not be content with the superficial structure that appears at first glance. And certainly not to be content with poor/inadequate data analysis in which the student sees the process as "run a few statistics tests, report the *p*-value, and call the analysis complete."

> Statistics is a tool to get where you want to go, but far too many view it either as an end for itself and the rest view it as a way of manipulating raw data in order to get a justification for what they want to do to begin with. Further, being able to start to quantify relationships and being able to quantify results does not mean that you are beginning to understand these, let alone being able to quantify anything like the risk involved
>
> *(Briggs, 2008)*.

I recently had to review a set of undergraduate honors research project proposals; they consistently had several weeks scheduled for data collection and one week for data analysis. Unfortunately, with only 1 week, these students will never get more than a superficial level of understanding of their data. In many of the cases of superficial analysis, I am more than willing to place a substantial part of the blame on the instructor. There is a substantial difference between a student who chooses to not to do a good data analysis and a student who does not know how to do a good data analysis. Unless students are taught how to perform an in-depth analysis, they will never perform one because they lack the knowledge. More importantly, they will lack the understanding to realize their analysis was superficial. If someone was taught the task as "do a *t*-test and report a *p*-value," then who is to blame for the lack of data analysis knowledge?

A goal of this book is to teach that data analysis is not just crunching numbers, but a way of thinking that works to reveal the underlying patterns and trends that allow a researcher to gain an understanding of the data and its connection to the research situation. I am content with students knowing when and why to use statistical tests, even if the test's internal logic is little more than a black box.

I expect many research methods instructors will be appalled at this book's contents. The heavy statistics-based researcher or a statistics instructor will be appalled at the statistical tests I left out or at the lack of rigorous discussion of many concepts. The instructor who touches on statistics in a research methods course will be appalled at the number of tests I include and the depth of the analysis. (Yes, I fully appreciate the inherent contradiction in these two sentences.) But I sincerely hope both groups appreciate my attempt at defining statistical tests as a part of data analysis—NOT as either its totality or its end—and my goal of teaching students to approach a data analysis with a mind-set of that they must analyze the data and not simply run a bunch of statistical tests.

With that said, here are some research issues this book will not address:

This book assumes the research methodology and data collection methods are valid. For instance, some examples discuss how to analyze the results of survey questions using Likert scales. Neither the design of the survey question or the developments of Likert items will be discussed; they are assumed to be valid.

This book assumes the data's reliability and validity. The reliability and validity of the data are research design questions that a well-designed study must consider up front, but they do not affect the data analysis per sec. Obviously, with poor quality data, the conclusions are questionable, but the analysis process does not change.

There are no step-by-step software instructions. There are several major statistical software packages and a researcher might use any one of them. With multiple packages, detailed-level software instructions would result in an overly long book with many pages irrelevant to any single reader. All the major software packages provide all of the basic tests covered in this book and there are essentially an infinite number of help sites and YouTube videos that explain the button-pushing aspects. Plus, the how-to is much more effectively taught one-on-one with an instructor than from a book.

The basic terminology used in research study design is used with minimal definition. For example, if the analysis differs for within subjects and between subject's designs, the discussion assumes the student already understands the concepts of within subjects and between subjects, since those must be understood before collecting data. Terminology relevant to a quantitative analysis will, of course, be full defined and explained. Also, there are extensive references to definitions and concepts.

There is no attempt to cover statistical proofs or deal with the edge cases of when a test does or does not apply. Readers desiring that level of understanding need a full statistics course. There are many places where I refer the researcher to a statistician. The complexities of much statistics or delving into more advanced tests may be relevant to the research, but are out of place here. This book is an introduction to data analysis, not an exhaustive data analysis tome.

This book focuses on the overall methodology and research mind-set for how to approach quantitative data analysis and how to use statistics tests as part of analyzing research data. It works to show that the goal of data analysis is to reveal the underlying patterns, trends, and relationships between the variables, and connecting those patterns, trends, and relationships to the data's contextual situation.

References

Briggs, W. (2008) The limits of statistics: black swans and randomness [Web log comment]. Retrieved from http://wmbriggs.com/blog/?p=204.

Rogers, J.L. (2010) The epistemology of mathematical and statistical modelling: a quiet methodological revolution. *American Psychologist*, 65 (1), 1–12.

About the Companion Website

This book is accompanied by a companion website:

   http://www.wiley.com/go/albers/quantitativedataanalysis

The website includes:

- Excel data sets for the chapter problems

1

Introduction

Basis of How All Quantitative Statistical Based Research

Any research study should have a solid design, properly collected data, and draw its conclusions on effectively analyzed data. All of which are nontrivial problems.

This is a book about performing quantitative data analysis. Unlike most research methods texts, which focus on creating a good design, the focus is on analyzing the data. It is not on how to design the study or collect the data; there many good sources that cover those aspects of research. Of course, poor designs or data collections lead to poor data that means the results of the analysis are useless. Instead, this book focuses on how to analyze the data.

The stereotypical linear view of a research study is shown in Figure 1.1a. Figure 1.1b expands on what is contained within the "analyze data" element. This book only works within that expansion; it focuses on how to analyze data from a study, rather than either how to perform the study or how to perform individual statistical tests.

The last two boxes of the expansion in Figure 1.1 "Make sense of the results" and "Determine the implications" are where performing a high-quality data analysis differs from someone simply crunching numbers.

A quantitative study is run to collect data and draw a numerical-based conclusion about that data. A conclusion that must reflect both the numerical analysis and the study context. Thus, data must be analyzed to *help* draw a study's conclusions. Unfortunately, even great data collected using a great design will be worthless unless the analysis was performed properly. The keyword in the sentence is *help* versus *give* the study's conclusions. The results of statistical tests are not the final conclusion for research data analysis. The researcher must study the test results, apply them to the situational context, and then draw conclusions that make sense. To support that process, this book

Introduction to Quantitative Data Analysis in the Behavioral and Social Sciences,
First Edition. Michael J. Albers.
© 2017 John Wiley & Sons, Inc. Published 2017 by John Wiley & Sons, Inc.
Companion website: www.wiley.com/go/albers/quantitativedataanalysis

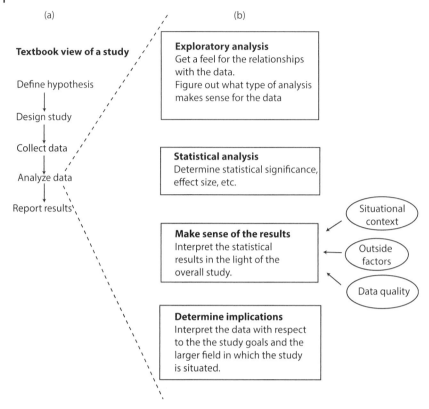

Figure 1.1 View of data analysis as situation within the overall study.

works to place the tests within the context of a problem and provide the background to connect a specific type of data with the appropriate test.

The outcome of any statistical analysis needs to be evaluated in terms of the research context and any conclusions drawn based on that context.

Consider this example of how this book approaches data analysis.

You are interested in which books are being checked out of a library. So, you gather data using many titles that fit within study-defined categories. For example, topical nonfiction or categories for fiction of a particular genre (historical, romance, etc).

At the end of the study's data collection, the analysis looks at the following:

- Graphs of checkouts by month of the various categories. Do the types of categories vary by day/week through the month? How do the numbers compare? Do the trends of checkouts for each category look the same or different?
- Run statistics on the daily/month checkouts of the book categories versus demographics of the people who checked them out (age, gender, frequency of

library use, etc.). Does age or gender matter for who checks out a romance versus a thriller. From this we can find whether there is a statistically significant difference (e.g., that older readers read more romance than younger readers).

Data Analysis, Not Statistical Analysis

Too many people believe if they can figure out how to run statistical software, then they know how to perform a quantitative data analysis. No! Statistics is only a single tool among many that are required for a data analysis. Likewise, the software is only a tool that provides an easy way to perform a statistical test. Knowing how to perform a *t*-test or an ANOVA is similar to knowing how to use styles and page layout in Word. Just because you know how to use styles does not make you a writer. It will not make you a good layout person if you do not know when and why to apply those styles. Neither the software nor the specific tests themselves are sufficient; necessary, yes, but sufficient, no! Run the wrong test, and the results are wrong. Fail to think through what the statistical test means to the situation and the overall study fails to have relevance.

It is important to understand that statistics is not data analysis. Learning how to use a software package to perform a *t*-test is relatively easy and quick. But good data analysis requires knowing when and why to perform a *t*-test; a much more different, and complex task. Especially for researchers in the social sciences, the goal is not to be a statistical expert, but to know how to analyze data. The goal is to be able to use statistical tests as part of the input required to interpret the study's data and draw valid conclusions from it. There is a wide range of statistical tests relevant to data analysis; some that every researcher should be able to perform and some that require the advice/help of a statistical expert. Good quantitative data analysis does not require a comprehensive knowledge of statistics, but, rather, knowing enough to know when it is time to ask for help and what questions to ask. Many times throughout the book, the comment to consult a statistician appears.

Figure 1.1 shows data analysis as one of five parts of a study; a part that deserves and often requires 20% of the full study's time. I recently had to review a set of undergraduate honors research project proposals; they consistently had several weeks scheduled for data collection, a couple of weeks for data clean up, and data analysis was done on Tuesday's. This type of time allocation is not uncommon for young researchers, probably based on a view that the analysis is just running a few *t*-tests and/or ANOVAs on the data and copying the test output into the study report. Unfortunately, with that sort of analysis, the researchers will never reach more than a superficial level of understanding of the data or be able to draw more than superficial conclusions from it.

The purpose of a quantitative research study is to gain an understanding of the research situation. Thus, the data analysis is the study; the study results come directly out of the analysis. It is not the collection and not the reporting; without the data analysis there is no reason to collect data and there is nothing of value to report.

Use dedicated statistical software

There are many dedicated statistical software programs (JMP, SPSS, R, Minitab) and many others. When you are doing data analysis, it is important to take the time to learn how to use one of these packages. All of them can perform all the standard statistical tests and the nonstandard tests, while important in their niche case, are not needed for most data analysis.

The one statistical source missing from the list is Microsoft Excel. This book uses Excel output on many examples, but it lacks the horsepower to really support data analysis. It is great for data entry of the collected data and for creating the graphs of the exploratory analysis. But, then, move on to a higher powered statistical analysis program.

What Statistics Does and Does Not Tells You

In statistics the word "significance" is often used to mean "statistical significance," which is the likelihood that the difference between the two groups is just an accident of sampling. A study's data analysis works to determine if the data points for two different groups are from the same population (a finding of not statistically significance) or if they are from different populations (a finding of statistically significance).

Every population has a mean and standard deviation. However, those values are typically not known by the researcher. Part of the study's goals is to determine them. If a study randomly selected members from the population in Table 1.1 any of those four groups could be picked.

If you take multiple samples from the same population, there will always be a difference between them. Table 1.1 shows the results of Excel calculating six random numbers that fit a normal distribution with a mean $= 10$ and standard deviation $= 2$. The numbers were generated randomly, but they could reflect the data from any number of studies: time to perform a task, interactions during an action, or, generally, anything that can be measured that gives a normal distribution. The important point here is that although they all come from the same population, each sample's mean and standard distribution is different.

Now put those numbers into a study. We have a study with two groups that are looking at the time to perform a task (faster is good). We have old way

Table 1.1 Random numbers generated with a normal distribution of mean = 10 and SD = 2.

	Trial 1	Trial 2	Trial 3	Trial 4
Data points	8.043	7.726	10.585	7.679
	7.284	7.374	9.743	12.432
	11.584	11.510	9.287	13.695
	9.735	11.842	9.102	8.922
	8.319	9.651	4.238	6.525
	8.326	11.849	11.193	11.959
Mean	8.882	9.992	9.025	10.202
SD	1.544	2.063	2.475	2.891

Because of the nature of random numbers and small sample sizes, each trial has a different mean and standard deviation, although they all come from the same population.

(performed by people that resulted in the times of trial 1: mean = 8.882, SD = 1.544) and a new way (performed by people that resulted in the times of trial 2: mean = 9.992, SD = 2.063). Because Table 1.1 presents data without context, a simple t-test (or ANOVA) to determine significance is all we can perform. If we simply looked at the mean and standard deviation numbers, it would seem trial 2 was worst, assuming we wanted fast task times. Yet, they are both random sets of numbers from mean = 10 and SD = 2. A proper statistical analysis would return a result of not statistically significance differences.

Within the context of a study, the data analysis requires moving beyond calculating a statistical value (whether a p-value or some of the results available with more complex statistical methods) and interpreting that statistical result with respect to the overall context. The research also has to determine if the difference is enough to have practical significance. From a practical viewpoint, there would be no reason to spend money to use the new way, since it is not any faster than the old way. If they had been significantly different, then the conclusions have to consider other factors (such as time and expense of the change) to determine if it is worthwhile. Quantitative data analysis is more than just finding statistical significance; it is connecting the results of the statistical analysis with the study's context and drawing practical conclusions.

Statistical significance is usually calculated as a "p-value," the probability that a difference of at least the same size would have arisen by chance, even if there really were no difference between the two populations. By social science convention, if $p < 0.05$ (i.e., below 5%), the difference is taken to be large enough to be "significant"; if not, then it is "not significant." In other words, if $p < 0.05$, then the two sets of data are declared to be from different populations. In the hard sciences, the p-value must be smaller.

Data is significant or not significant. There is no maybe. Maybe . . .

Research reports often contain sentences such as "The *p*-value of 0.073 shows the results are trending toward significance." Researchers have long running debate about this type of wording. The basis of the argument is that a result is yes/no: it is either significant or it is not significant.

However, the definition of significance as $p = 0.05$ is itself arbitrary and is only a long-standing convention in the social sciences.

More importantly, thinking in terms of a statistical yes/no ignores the study context. The effect size (loosely defined as the practical significance) is more important in the final result than the statistical significance. If a study fails to show statistical significance—in this case, $p = 0.073$—but has a large effect size, the results are much strong than one with the same *p*-value (or even $p = 0.04$ that is significant) but a small effect size.

Both *p*-value and the effect size should be reported in a study's report.

Some researchers try to skip the statistical analysis and observe and draw conclusions directly from the data. Looking at the data in Table 1.1 without a statistical analysis might conclude that trials 1 and 4 were from different populations, but this is not the case. If this was a study to determine if a current practice should change, lots of money could be spent on a change that will have no real affect. Without a proper analysis, the researcher can fall prey to confirmation bias. A confirmation bias occurs when the data to support a desired claim is specifically looked for and data that refutes the claim is ignored. A simple example is politicians from opposing parties using the same report to claim their agenda is right and the opposing agenda is wrong. Yet, both are cherry picking data from the report to make that claim.

Although statistics can tell you the data are from different populations (there is a statistical significant difference), it does not tell you the *why*." Yet, the purpose of a research study is to uncover the why, not just the proof of a difference. Thus, a statistical analysis itself is not the answer to a hypothesis. A research study needs to move beyond the statistics and answer questions of why and what really happened to give these results, and then move past those questions to figure out what the answers mean in the study's context. That is the outcome of a good data analysis.

Data Analysis Focuses on Testing a Hypothesis

Early in a study's design, a set of hypothesis is created. Then the data needed to test those hypotheses are defined and eventually collected. At the same time the

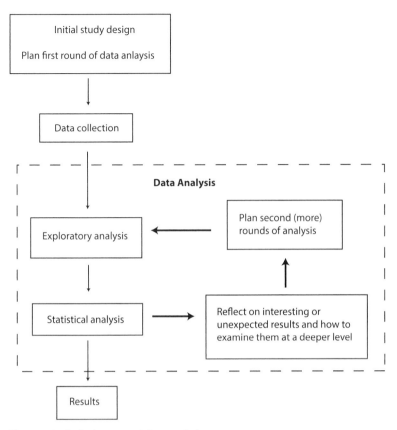

Figure 1.2 Cyclical nature of data analysis.

needed data is defined, the analysis tests to be performed on the data should be defined.

Good data analysis knows from the start how the first round of analysis will be performed; that was defined early in the study design. The second (and other) rounds of analysis that each drill deeper and explore interesting relationships found in the previous rounds. Obviously these cannot be defined until you are engaged in the analysis, but understanding how to pursue them distinguishes a poor from a good researcher (Figure 1.2).

Poor data analysis often just collects data and then runs all possible combinations of data, most of which make no sense to even test against each other. The goal of the data analysis has been shifted from understanding the data within the study to "finding significance . . . any significance." Unfortunately, with significance defined as $p < 0.05$ that means 5% of the combinations may show significance when it does not exist.

The best, most well-designed study is worthless if the data analysis is inadequate. Focus the analysis on a small number of well-conceived hypotheses rather than blindly performing different statistical tests on all variable pairs and ending up with 5% of your results being significant at the 0.05 level.

Quantitative Versus Qualitative Research

Research methodology textbooks call for qualitative research to get a view of the big picture and then to use quantitative studies to examine the details. Both quantitative and qualitative have their strong and weak points, and a good research agenda requires both. That implies that a researcher needs to understand both. A problem occurs—similar to the "when you have a hammer, everything looks like a nail" problem—when researchers lack training in one, typically quantitative, and try to use a single approach for all research problems.

Quantitative research is a methodical process. Too many people with qualitative research experience—or who lack quantitative research experience— look at a situation, point out all of the interacting factors, and despair (or claim an impossibility) of figuring out the situation.

The social sciences work with large complex systems with numerous variables interacting in subtle ways. The reality is that it requires many studies with each looking at the situation in slightly different ways and with each study exposing new questions and new relationships. Qualitative research can define the variables of interest and provide preliminary insight into which variable interact and help define the hypothesis, but it requires quantitative research to clearly understand how those variables interact.

> It is one thing to declare confidently that causal chains exist in the world out there. However, it is quite another thing to find out what they are. Causal processes are not obvious. They hide in situations of complexity, in which effects may have been produced by several different causes acting together. When investigated, they will reluctantly shed one layer of explanation at a time, but only to reveal another deeper level of complexity beneath.
>
> *(Marsh and Elliott, 2008, p. 239)*

As its strongest point, quantitative data analysis gets at the deep structure of the data. High-quality quantitative data analysis exposes a deep structure and researchers should never be content with the superficial structure that appears at first glance. And certainly should not to be content with poor/inadequate data analysis where the analysis process is seen as running a few statistics tests, reporting the p-value, and calling that a data analysis.

It is true that no social science study will be able to obtain the clear results with the hard numbers obtained in the physical sciences. The cause and effect relationships of the laws in the physical sciences do not exist. Instead, when people enter into the research equations, there are at best probabilistic relationships. Nothing can be clearly predicted. But to simply refuse to undertake a study because of too many interactions is poor research, as is deciding to only undertake qualitative research. Qualitative research can build up the big picture and show the existence of relationships. As a result, it reveals the areas where we can best apply quantitative approaches. It is with quantitative approaches that we can fully get at the underlying relationships within the data and, in the end, it is those relationships that contain the deep understanding of the overall complexities of the situation (Albers, 2010). Developing that deep understanding is the fundamental goal of a research agenda.

What the Book Covers and What It Does Not Cover

Every quantitative research study collects some type of data that gets reduced to numbers and must be analyzed to help draw the study's conclusions. A great study design is useless unless the data are properly analyzed. What I have found is that most textbooks fall into one of these categories.

- Research method textbooks explain how to create and execute a study, but typically are very light on how to analyze the data. They are excellent on explaining methods of setting up the study and collecting the data, but not on the methods to analyze data after it has been collected.
The data analysis chapters of many research textbooks are little more than a brief explanation of various statistical tests. As a result, reader can come away thinking the important questions to ask are "How do I run a chi-square?" "What is the best procedure, a Kruskal–Wallis test or a standard ANOVA?" and "Let me tell you about my data, and you can tell me what procedure to run" (Rogers, 2010, p. 8). These are the wrong questions to be asking at the beginning of the data analysis. Rather, data analysis needs to be addressed along the lines of "what relationships do I need to understand?" and "what do these results tell me about the research context?"
- Statistics textbooks explain how to perform statistical tests. The tests are explained in an acontextual manner and in rigorous statistical terms. They explain *how* to perform a test, but, from a research standpoint, the equally important question of *when* and *why* to perform a test gets short shrift. Likewise, statistics textbooks do not explain the need to connect the statistical results to the research context.

This book differs from these two categories because it is focused on explaining how to analyze data from a study, rather than how to perform the study or how to perform individual statistical tests. Early in the Chapter 1 said "data that must then be analyzed to *help* draw the study's conclusions." The keyword in the sentence is *help* versus *give* the study's conclusions. The results of statistical tests are not the final conclusion for research data analysis. The researcher must study the test results, apply them to the situational context, and then draw conclusion that make sense.

The importance of data analysis is clearly summed up within this quote by Phillips.

> In moving from data collection to data analysis, we confront a variety of complex mathematical procedures. Thus it is essential not to lose sight of the theoretical aspects of the research process as we become involved in those procedures. The scientific method may be seen as a continuing chain. If our theoretical definition of the problem becomes a series of weak links, the entire chain is weakened. The reverse holds true as well: We do not want our procedures for analyzing data to become weak links in the scientific method. This requires that we learn to make use of the best methods available. Further, since we wish to strengthen the entire chain, we must also learn the limitations of existing analytic tools and be open to ways of solving the problems involved.
>
> *(Phillips, 1985, p. 386)*

The bulk of the book is examples that walk through a data analysis. They explain the exploratory analysis of getting a good feel for the data and then explain why various statistical tests are performed on the data and how to interpret them.

The examples in this book will not address whether the variables from the data collection make sense or even how the data are collected. These are issues that must be addressed early in the experimental design and clearly affect the overall results, but they are not a data analysis concern. Granted, if the categories make no sense, then the results make no sense. But that is a fault of a poorly designed study, not a fault of poorly performed data analysis.

It is, unfortunately, very easy to have a well-designed study with good data that suffers from poor data analysis or a poorly designed study with bad data that has great data analysis (Table 1.2). In either case, results are useless and may even be misleading. All research studies should strive to be in the lower right quadrant.

Book Structure

This book is striving to explain the when, why, and what for, rather than the button pushing how to.

Table 1.2 Four quadrants of study design and data analysis.

		Data analysis	
		Poor	**Good**
Study design	Poor	Poorly designed study, poorly performed data analysis. **Bad data that is poorly analyzed. The study is a complete failure.**	Poorly designed study, well-performed data analysis. **Bad data that is properly analyzed. Unfortunately, since it started with bad data, the results are useless.**
	Good	Well-designed study, poorly performed data analysis. **Good data, but with questionable results. With proper analysis, a good study can be salvaged.**	Well-designed study, well performed data analysis. **A study with valid results.**

To support the process of how-to-do data analysis, this book works to place the tests within the context of a problem and provide the background to connect a specific type of data with the appropriate test. The work is placed within long examples and the entire process of data analysis is covered in a contextualized manner. It looks at the data analysis from different viewpoints and using different tests to explain how and when to apply different analysis methods.

The ideas and concepts of data analysis form a highly interconnected web. Not understanding a concept makes understanding its application difficult. On the other hand, explaining a concept repeatedly makes a text difficult to read for people who do understand it. To help with the problem, the book uses extensive cross-reference concepts to the page where they are discussed in detail.

The goal of this book is to emphasize that data analysis is not just crunching numbers, but a process of revealing the underlying patterns and trends that allow a researcher to gain an understanding of the data and its connection to the research situation.

References

Albers, M.J. (2010) Usability and information relationships: considering content relationships when testing complex information. In: M.J. Albers and B. Still (Eds) *Usability of Complex Information Systems: Evaluation of User Interaction*, Boca Raton, FL: CRC Press.

Marsh, C. and Elliott, J. (2008) *Exploring Data: An Introduction to Data Analysis for Social Scientists*, 2nd edn, Cambridge: Polity.

Phillips, B. (1985) *Sociological Research Methods: An Introduction*. Homewood, IL: Dorsey.

Rogers, J.L. (2010) The epistemology of mathematical and statistical modeling: a quiet methodological revolution. *American Psychologist*, **65 (1)**, 1–12.

Part I

Data Analysis Approaches

This section provides a basic introduction to the concepts of quantitative data analysis and the approaches that a researcher should use.

Introduction to Quantitative Data Analysis in the Behavioral and Social Sciences,
First Edition. Michael J. Albers.
© 2017 John Wiley & Sons, Inc. Published 2017 by John Wiley & Sons, Inc.
Companion website: www.wiley.com/go/albers/quantitativedataanalysis

2

Statistics Terminology

This section provides a discussion of the main statistical terms used for the examples and relates how they are relevant to data analysis. It gives basic definitions, but does assume the reader has seen the terms before.

Statistically Testing a Hypothesis

It is with quantitative approaches that we can fully get at the underlying relationships within the data and, in the end, it is those relationships that contain the deep understanding of the overall complexities of the situation design requires the hypothesis to be written in a form to support null-hypothesis testing. In other words, the hypothesis is often written as "there will be no changes in the dependent variable observed with changes in the independent variable." The null hypothesis is a statement that exists solely to be falsified by the data set; the study wants to reject the null hypothesis. Statistical significance is a finding of the statistical test that the data set is unlikely to have occurred by chance *if the null hypothesis is true*. One characteristic of the statistical test is a black and white result: reject or do not reject.

A statistical test examines whether the control and experimental group data are from the same population. The null hypothesis states "there will be no change in the mile run time when people run 45 min per day for 2 weeks." We have a control group that did not exercise for 2 weeks and the experimental group that did. The null hypothesis claims that the data sets are not different (exercising will not change the mile run time), which is typically defined as obtaining a p-value of $p > 0.05$. In other words, the null hypothesis claims the mile run times at the end of the 2 week will be the same for the control and experimental group. Not a likely outcome and, in fact, one that the study wants to show is false. The study wants to show that exercising will improve the mile run time—it wants to reject the null hypothesis.

Introduction to Quantitative Data Analysis in the Behavioral and Social Sciences,
First Edition. Michael J. Albers.
© 2017 John Wiley & Sons, Inc. Published 2017 by John Wiley & Sons, Inc.
Companion website: www.wiley.com/go/albers/quantitativedataanalysis

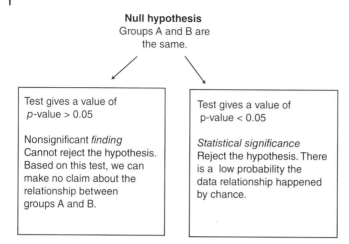

Null hypothesis
Groups A and B are
the same.

Test gives a value of
p-value > 0.05

Nonsignificant *finding*
Cannot reject the hypothesis.
Based on this test, we can
make no claim about the
relationship between
groups A and B.

Test gives a value of
p-value < 0.05

Statistical significance
Reject the hypothesis. There
is a low probability the
data relationship happened
by chance.

Figure 2.1 Testing a hypothesis. We can reject a hypothesis and claim the changes between groups A and B is not by chance if the *p*-value is less than 0.05. But we can make no claim in the *p*-value greater than 0.05.

Statistical tests are run on the data and, hopefully, they show that the hypothesis is rejected. When we reject a hypothesis based on *p*-value <0.05, what we are actually claiming is that if the null hypothesis is true, then there is a low probability the differences in the control and experimental groups happened by chance (Figure 2.1).

A closer examination of that last sentence reveals that the *p*-value is really measuring. Remember the different sets of means calculated and graphed in Table 1.1. Any two of those numbers could fit the data for the control and experimental group. The *p*-value shows where the experimental group falls on that graph and, we strongly suspect, the values will have greatly improved and no longer be in the same population.

The standard statistical tests covered in most textbooks (and that will be covered in this one) are tests focused on null-hypothesis testing, because they are testing a study null hypothesis to show it to rejected or not rejected by the test. A statistical test may reject a null hypothesis, but that is not the endpoint of the data analysis. The researcher then needs to interpret what the rejection of the null hypothesis means in practical terms and in terms of the rest of the study data.

One caution about rejecting a hypothesis based on *p*-value alone. Obtaining a reject result means the hypothesis is not supported. In common terms, the hypothesis is worded so that it predicts no change (smoking does not cause lung cancer) and when the resulting statistical test shows it is rejected, the researcher concludes the alternative hypothesis that smoking does cause lung cancer. However, the researcher must examine the data further to ensure that conclusion is warranted. The problem is that a statistical test does not and cannot distinguish between a cause–effect relationship (which the researcher wants to show) and a correlation relationship (which while interesting, requires more digging to reveal the underlying cause–effect relationship).

For example, we use ironworkers union and teachers union as the independent variable and measure the member's arm strength. We would probably find a strong correlation, with the ironworkers being stronger. However, there is no cause–effect relationship. The act of joining the ironworkers union does not make a person stronger or joining the teacher's union make a person weaker. It is the underlying nature of the job, not the union membership, that contains the cause–effect variable.

The statistical tests reveal the two groups are different, and we are assuming it is the because of the change in the independent variable caused the change in the dependent variable. The strength of that claim depends on how well the study handled the controlled variables. The researcher needs to consider the possibility that a third variable (known or unknown) was changing independent of the independent variable and it was the third variable that caused the dependent variable to change. The study design and an in-depth data analysis work to either eliminate or tease out those third variables.

A study needs to define its data measures to ensure the validity and reliability of the study, but those factors are independent of the statistical test. They affect the interpretation of the data and any practicality of the study, but not the performance of the statistical test itself.

A hypothesis must be phrased in a testable manner

Every hypothesis must be phrased so that it can be tested. A hypothesis of "does doing grammar exercises improve writing ability?" is not worded such that it can be nullified by a statistical test. The wording should be more like "performing daily grammar exercises will not enhance writing ability." Then a statistical test can be performed on a measure collected from the grammar exercises and on the student's writing ability. A researcher's hope is that the test will show statistical significance ($p < 0.05$), so he can conclude that the hypothesis is not supported, which supports the alternative idea that grammar exercise does improve writing.

Limitations of Null-Hypothesis Testing

When the hypothesis is rejected—typically the desired outcome of a test—a study can claim support for the idea being researched. However, failing to reject a null hypothesis does not imply support for the hypothesis. In other words, a *p*-value that indicates no statistical significance does not indicate that the hypothesis is ipso facto correct.

Failing to reject results ($p > 0.05$) means the data set is compatible with the null hypothesis. It does not imply that the null hypothesis is right or wrong, but that the data cannot determine a right/wrong status. *Thus, "do not reject hypothesis" is not the same as "accept the hypothesis."* Rejecting a hypothesis means the test determined the data sets differed by more than chance; it does not prove the alternative is true, namely that we should accept the hypothesis. This difference can be subtle and is confused by many researchers.

The fundamental problem is that null-hypothesis statistical testing does not tell us what we really want to know (Kline, 2004). What we want to know is whether the null hypothesis is true given the data. Granted, we typically want to reject the null hypothesis, but that is equivalent to asking if it the hypothesis is true given the data. In contrast, null-hypothesis statistical testing actually gives the inverse. It does not give what we want to know—is the null hypothesis true given the data—instead it indicates the probability that if the null hypothesis were true then the data could have been obtained by chance (Cohen, 1994).

When the test fails to reject the null hypothesis that rejection implies *inconclusive* results. It does not mean that there is no relationship. The preceding sentences can be summed up as failure to reject the null hypothesis does not mean an acceptance of the null hypothesis. There may have been too little data, the data collection method was too insensitive, the data was too noisy, there was a chance selection of the wrong members in the participant samples, etc.

Thus, a study result can claim A and B did show a statistical significance, however, if the test fails to show statistical significance, it cannot claim that A and B do not have a relationship. Instead, it can only claim the results are inconclusive. If we reject the hypothesis, we can claim that giving grammar exercises proves writing ability, but if we cannot reject the hypothesis we cannot claim they do not. Perhaps the grammar exercises were poorly constructed, too trivial, not enough of them, did not relate well to the student's grammar errors, etc.

It is important to always remember than the *p*-value describes the likelihood of observing certain data if the null hypothesis is true. A small *p*-value is the evidence against a null hypothesis. It says nothing about whether the null hypothesis is correct, only if there is significant evidence to reject it—we can make a claim that the control and experimental groups are from different populations.

In the end, the main ideas of the preceding paragraphs need to be understood by quantitative researchers, although the nuances can be left to philosophers of science. Most researchers need to keep in mind the real meaning of null-hypothesis testing as they contemplate the results of the statistical tests and relate them back to the study's context. What does it mean to have a high probability that the relationship between two variables would not arise by chance? Does it make sense with respect to the study design? What does this mean based on the goals of the study?

Statistical Significance and *p*-Value

Generally, to be publishable, quantitative research findings are expected to show statistical significance. Statistical significance is expressed by the *p*-value of the statistical function.

Statistic tests return a probability called the *p*-value that defines whether or not the test has found the differences between the two groups to be statistically significant (Figure 2.2). A parametric statistical test looks at the distribution of the data points and determines whether the two sets of data are both parts of the same normal distribution population. The *p*-value is the chance of obtaining the result if the null hypothesis is true. Or it can also be considered as the probability that the second set of data comes from the same population as the first set of data.

Formally, the *p*-value is the probability of obtaining a test statistic at least as extreme as the one that was actually observed, assuming that the null hypothesis is true. Thus, the lower the *p*-value, the stronger the evidence that the null hypothesis is false. Traditionally, in the social sciences, the null hypothesis is

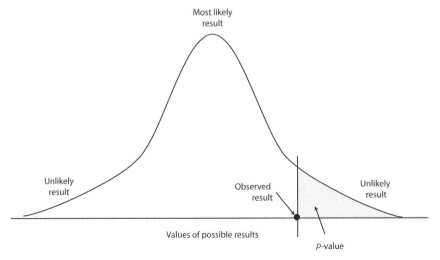

Figure 2.2 *p*-Value of an observed result.

rejected if the p-value is below 0.05, which means there is only a 5% chance that the observed results would be seen if there were no real effect. In the physical sciences, a smaller p-value may be required.

A p-value of less than 0.05 means there is only a 5% chance that the observed results would be seen if there were no real effect. Meaning, the probability of getting the two sets of data if the null hypothesis—no difference between groups—is true.

When the p-value is less than 0.05, then we declare the null hypothesis is not supported and the two groups are from different populations. However, the p-value actually says the opposite. A misconception that consistently—and erroneously— happens in many study reports.

> A recent popular book on issues involving science, for example, states a commonly held misperception about the meaning of statistical significance at the 0.05 level: "This means that it is 95 percent certain that the observed difference between groups, or sets of samples, is real and could not have arisen by chance." That interpretation commits an egregious logical error (technical term: "transposed conditional"): confusing the odds of getting a result (if a hypothesis is true) with the odds favoring the hypothesis if you observe that result. A well-fed dog may seldom bark, but observing the rare bark does not imply that the dog is hungry. A dog may bark 5 percent of the time even if it is well-fed all of the time
> *(Siegfried, 2010, p. 26).*

Using the p-Value

A study has two groups (control and experimental) that may or may not be from the same population. The basic purpose of any statistical test is to answer that question about whether they are from the same or different populations. Table 1.1 showed four sets of random numbers all from the same population. Although they are from the same population, they show a range of mean and standard deviation values; any of two of these might be the data collected for the control and experimental groups. The statistical test is performed and p-value is used to make a claim of statistical significance. The research then needs to think about the big picture and make sure that claim makes sense with respect to the entire study.

A common misconception is that a p-value is the probability that the studies values occurred by random chance (such as the differences in Table 1.1). Not quite. Its real meaning exists in a bigger view: The p-value puts a value on how often results at least as extreme as those observed would be found if the study were repeated an infinite number of times with subjects from the same population and *only* random chance caused the data variation.

An additional problem to handle when reporting statistical significance is the problem of testing many items, such as individually performing a *t*-test on each questions of a survey taken between two groups. If the survey had 20–30 questions, there is a high probability at least one will show a result of being statistical significant when it is not. Loosely interpreted, a *p*-value of 0.05 means 1 in 20 tests will show an unjustified significance. In studies with large numbers of tests, such as genetics where hundreds or thousands of genes may be compared, the test design must address this issue. Even in survey analysis, any significance for an individual question must be balanced against the entire analysis. The researcher must ask if it makes sense for just that one question out of 40 to show significance or if it is a random fluctuation. Making this call requires considering the confidence intervals and effect size, as well as the nature of the survey question compared to the other questions.

Test Statistic and *p*-Value

A statistical test returns a test statistic. Formally, a test statistic is a function of all the observations that summarizes the data into a single number. If you calculate a statistical test by hand, the test statistic is the resulting number. That test statistic is used to calculate the *p*-value. A table (Figure 2.3) is used to convert the test statistic into a *p*-value. With software-based statistical analysis, the test statistic is no longer of great importance; however, it is commonly reported in the study's write up.

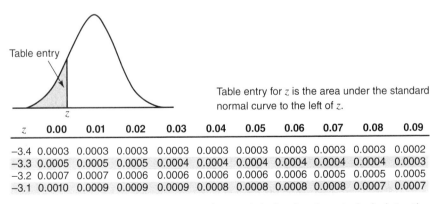

Table entry for z is the area under the standard normal curve to the left of z.

z	0.00	0.01	0.02	0.03	0.04	0.05	0.06	0.07	0.08	0.09
−3.4	0.0003	0.0003	0.0003	0.0003	0.0003	0.0003	0.0003	0.0003	0.0003	0.0002
−3.3	0.0005	0.0005	0.0005	0.0004	0.0004	0.0004	0.0004	0.0004	0.0004	0.0003
−3.2	0.0007	0.0007	0.0006	0.0006	0.0006	0.0006	0.0006	0.0005	0.0005	0.0005
−3.1	0.0010	0.0009	0.0009	0.0009	0.0008	0.0008	0.0008	0.0008	0.0007	0.0007

Figure 2.3 Test statistic table for a *t*-test. Working statistics longhand required calculating the test statistic z and then using a table to look up the *p*-value. Thankfully, software handles this aspect of the calculation.

Resampling to get the *p*-value

p-Values are not actually calculated this way, but this example gives a good illustration of what the *p*-value actually measures.

One way to think about the *p*-value (and the confidence interval) is to consider the results of resampling the data many times.

The null hypothesis for a study would state that there is no difference between the control and experimental groups; in other words, they both come from the same population.

A study collects 500 data points each for a control and an experimental condition, for a total of 1000 data points. If the null hypothesis is true, then all 1000 of these data points come from the same population. So, working on that assumption, we are going to randomly pick 100 points out of the 1000 (without regard for whether control or experimental group) and plot the mean for those 100 points. Clearly, this mean of 100 random points does not give useful data, but we will repeat the random sampling many times. (This is sampling with replacement since every time a sample of 100 points is selected, all 1000 data points are used. Whether a point was previously picked is not relevant to the sample selection.)

The means that we calculate will differ from the mean of the 1000 points; some means will be higher and some will be lower. Although a very low probability, there is a nonzero chance that the 100 points picked were the 100 smallest/largest in the data set.

After enough samples, the means of each of the samples will plot into a normal distribution (Figure 2.4). The height of the curve is equal to how many

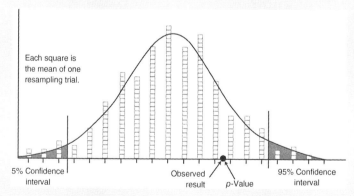

Figure 2.4 Plot of the means of the data resampling. The 5 and 95% points give the confidence intervals. The plotted point of the mean of the experimental sample is its *p*-value.

times that specific mean was calculated. We expect the sample mean to be close to the population mean and to cluster around that spot, giving the curve peak at approximately the population mean. Interestingly, this population mean may differ from the mean of the 1000 points.

Now, we calculate and plot the position of the mean of the experimental data (using all 500 points). This mean is equal to the *p*-value. If it is less than the 5% or greater than the 95% lines of the graph (two standard deviations from the mean developed from the resampling), then the results are considered significant.

It is important to notice that during the random sampling, there were some calculated means that fell outside of the 5 and 95% range. This is normal and expected. In a small data set, a researcher might be tempted to toss them as outliers. There is also a chance that the experimental data, because of random chance, fell outside the 5 and 95% range. In other words, this is also a visual representation that although we can say the null hypothesis was rejected, there is still a chance that the null hypothesis is true and it is the random fluctuation of the data itself that caused the rejection.

p-Values cannot be compared

The *p*-value cannot be used to determine which of multiple experimental groups is better or worse than other groups. The *p*-value calculation is only meaningful for the data used to calculate it; it is meaningless to compare it to other *p*-values.

A company is looking to replace a computer system. Two new interface designs (B and C) and the old design (A) are tested for speed of data entry. *t*-Tests are used to compare the times. The times for both B and C are faster and the *p*-values showed they were statistically significant as compared with A:

A–B $p = 0.034$
A–C $p = 0.001$

A common misconception is that this result supports a claim that C is better than B because the *p*-value is smaller. *This is false*; the *p*-value has no bearing on whether B or C is better. The study can conclude that both B and C were statistically significant from A, but cannot say anything about a direct comparison of B and C.

An additional *t*-test comparing B and C may conclude they have a statistically significant difference, but that still says nothing about practical significance of which is better.

p-Value is not an event's probability

Another source of confusion with *p*-values is that some people want to apply the value to the data used to calculate it. In other words, they confuse a *p*-value that comes from a test (let us say $p = 0.65$) with a probability of 65% of an event occurring. These are two very different things. The *p*-value says nothing about individual data points.

I heard the misconception explained as "we tested 100 people and with $p = 0.65$, we know that 65 people were correct and, thus, 35 were not correct." This logic was extended to a drug test where 35 people had a false positive test. No. The logic would be correct if there was a 65% probability of an event occurring, but that is not what a *p*-value means.

The *p*-value must always be considered a group value. We have 100 people in our experimental group and since $p = 0.65$, we conclude that as a group those 100 are not different from the control group. It says nothing about probabilities of any event or any individuals in that group.

Maybe *p*-Value Should Not Be Used

This book uses the standard of the *p*-value as part of null-hypothesis testing. However, within the statistical community, there continues a long-term discussion on whether *p*-values should be used (Jones and Tukey, 2000).

The problem is not about *p*-values, per se, but about how they are misunderstood and misused (Nuzzo, 2015). Some journals are now banning the use *p*-values in reporting results (Trafimow and Marks, 2015). The alternative to null-hypothesis testing is using Bayesian statistics.

Sample Size and Statistical Significance

Sample size has a strong influence on how easily a test will show statistical significance. Social science data often suffers from small data sets that can make it difficult to obtain statistical significance.

A large or a small sample size can affect the study results.

- Large data sets, such as those used in clinical trials, may show significance for very small differences.
- Small data sets with relatively large standard deviations rarely show significance because of the lack of power of the test.

Change in *p*-value with increased data set size

Interestingly, repeating the same data can also cause the test to show significance.

Table 2.1 takes the same numbers and increases their value. Excel generated 7 random numbers between 0 and 9, listed in the first two columns. A *t*-test on these numbers shows a nonsignificant *p*-value as would be expected, since they are random numbers from a uniform distribution.

Table 2.1 Repeating seven random numbers and change in *t*-test *p*-value.

5	1
8	7
4	1
1	0
6	2
1	8
8	7
$p =$	0.569

5	1
8	7
4	1
1	0
6	2
1	8
8	7
5	1
8	7
4	1
1	0
6	2
1	8
8	7
$p =$	0.397

Columns 3 and 4 show the same numbers repeated; note how rows 8–14 duplicate rows 1–7. Also, note how the *p*-value, although still nonsignificant, is smaller. Continuing in this manner and repeating the seven value 12 times (total of 84 numbers in each column) gives a *p*-value equal to 0.031, which is significant.

Remember the numbers randomly selected. Of course, data collection would never reuse the same numbers like this, but it does highlight the sensitivity of large data sets.

Confidence Intervals

Confidence intervals are mathematically related to, but are NOT the same as the *p*-value. This can be confusing since both often use similar values of 0.90 or 0.95. The shaded end tails of Figure 2.4 are the confidence intervals.

The confidence level describes the uncertainty associated with a sampling method. It gives an estimated range of values that are likely to include the variable of interest, with the estimate being calculated based on the sample data. In other words, the confidence interval describes the uncertainty associated with how the data was collected. A study can only report the study mean, which is assumed (hoped) to be close to the reality—the population mean. Confidence intervals give a method of quantifying that hope. A 90% confidence interval means that if a study was ran many times, that 90% of the time the study confidence interval would include the population mean. A 95% confidence level means that 95% of the intervals would include the population mean; and so on.

Confidence intervals give a numeric range that describes the probability that the real population value will be within that range. For example, a study on improving reading skills finds an average improvement of 12 words per minute with a 90% confidence interval of 5–19. This means that although the study found a mean improvement of 12, the real improvement is probably (we are 90% confident) between 5 and 19 words per minute. If the study were run many more times, the average of all of the studies would converge on the real population mean. Of course, cost and practicality prevents running a research study many times.

Confidence interval should be reported as part of the statistical analysis. Typical choices of confidence intervals in the social sciences are the 90 and 95% levels. These levels correspond to percentages of the area of the normal curve. For example, a 95% confidence interval covers 95% of the normal curve—the probability of observing a value outside of this area is less than 0.05.

Confidence interval size is strongly dependent on sample size. As a sample gets larger, the confidence interval gets smaller; eventually the sample becomes large enough that its difference from the population is trivial. A study needs to use a large enough sample to make the confidence interval small enough to be of practical value. The connection with sample size helps to keep the study's focus on the effect size and practical significance, rather than just the statistical significance reported with the *p*-value. It is also the reason that studies with small sample sizes are not considered reliable; their large confidence intervals give too much spread to have strong confidence in the study value.

Although many publications suggest (almost mandate) the use of 95% confidence intervals, the acceptable values can be less. Social science and applied research must balance the priority given to type I and type II errors that may require using a lower confidence interval. Otherwise, with noisy data, finding significant results would be almost impossible.

What Confidence Interval Actually Means

Many people confuse what the confidence interval actually reports. They think it means, using the earlier example, that a mean value of 12 gives a 90% chance that the population's mean falls between 5 and 19. However, the population mean is a constant, regardless of what the study found, the population mean does not change. The confidence interval is actually the reverse of how many people interpret it. The confidence interval provides a measure of the uncertainty associated with the data collection. An incorrect interpretation would be that there is a 90% chance that the population mean falls between 5 and 19. What it really says is that if the study were rerun many times, 90% of those times the study results would fall within the confidence interval that contain the real population mean. On any specific study run (a researcher typically only has one set of data), there is no way of knowing if the study is within the 90% spanned by the confidence internal or within 10% outside of it.

Effect Size

The idea of practical significance and the need for the data analysis to connect to it has been emphasized multiple times. At one level, the researcher has to reflect on the study details and how they connect to the statistical analysis. However, statistical significance does not tell you the most important thing: the size of the effect (Coe, 2002).

But there is another factor, effect size, to consider when thinking about the practical significance. Unfortunately, most textbooks explain the statistics, but fail to mention the issue of effect size. Effect size gives a number that describes the size of the difference between two groups, and could be considered as a measure of the significance of the difference between the groups. The effect size

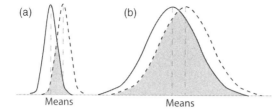

Figure 2.5 Effect size. (a) There is a strong effect since there is little overlap between the two curves. (b) The effect would be barely noticeable because of the overlap. In both a and b, the curves have the same mean, but differences in the standard deviations cause the differences in curve overlap.

is a term that can be considered, more or less, as an equivalent term to practical significance. Calculating and reporting a large effect size of the data sample strengthens the argument that that the finding of statistical significance has real meaning. It helps support the "so what?" aspects of a study's conclusions.

For example, if a study examined the quality of writing versus the time of day when the student wrote the text, we might find that mid-morning writing sample was better than late afternoon writing sample and that difference was statistically significant. Then, to consider the practical significance of the findings, we need to ask: "How much difference would it make if student's always write mid-morning versus in the late afternoon?" The effect size provides an answer to that question. If the effect size is small, then there would be no reason to try and get student's to change when they write. On the other hand, a large effect size could make noticeable change in overall student grades.

In Figure 2.5, there are two sets of curves—perhaps they are the plots of two different studies of student writing scores from mid-morning and late afternoon. Both sets may have the same difference in their mean values, but clearly the standard deviation is different. More important, from a data analysis view, is to think about the amount of overlap between the two curves. A *t*-test may reveal that the difference of both sets of data is statistical significant, but it does not reveal (or even consider at any level) the amount of overlap between the curves. Yet the amount of overlap gives an indication of how strong the findings really are. In curve B, the overlap is so substantial that there is a very small effect.

Calculating Effect Size

There are different effect size equations that are designed to handle whether the two groups have different standard deviations. There are also effect size equations for multiple groups—when the proper test would be an ANOVA. They are calculated differently, but they are interpreted similar to Cohen's *d*, which is the typical one used with two groups.

Formally, effect size is a standardized, scale-free measure. There are many different methods of calculating effect size, but a common one is Cohen's *d*.

Statistics software may calculate it or there are many online calculators for effect size.

$$\text{Effect size} = \frac{[\text{mean of experimental group}] - [\text{mean of control group}]}{SD_{pooled}}$$

$$SD_{pooled} = \sqrt{\frac{SD_{exp}^2 + SD_{control}^2}{2}}$$

The standard deviation term is the standard deviation of the population of the two groups. This is rarely known and must be estimated using the a pooled SD of the two values.

Depending on if the experimental group mean is larger or smaller than the control group mean, the equation could yield a negative number. However, the absolute value is the same and this is what to use in evaluating the effect size.

• An effect size of 0 means the experimental and control groups performed the same.
• A negative effect size means the control group performed better than the experimental group.
• A positive effect size means the experimental group performed better than the control group.

Since an effect size of 0 means no difference, the closer to either −1 or 1 the more effective was the study treatment.

Table 2.2 shows how to evaluate the effect size results from Cohen's *d*.

Tables 2.3 and 2.4 show how the value of Cohen's *d*—the effect size—changes when the either the standard deviation changes or the group means change.

A potential problem with interpreting effect size is that the Cohen's *d* assumes a normal data distribution for both the control and experimental groups. If this assumption is violated, it is best to consult a statistician when calculating effect size.

Table 2.2 Interpretation of Cohen's *d* results.

Cohen's *d* effect size	Interpretation of value
$0 < d < 0.2$	No effect (or minimal)
$0.2 < d < 0.5$	Small
$0.5 < d < 0.8$	Medium
$0.8 < d$	Large

Table 2.3 Changes in Cohen's *d* result as the standard deviations change.

SD with ave = 42	SD with ave = 45	Effect size	Interpretation
3	2	1.177	Large
3	4	0.849	Large
3	6	0.632	Medium
3	8	0.497	Small

Table 2.4 Changes in Cohen's *d* result as the average values change.

Group 1 mean	Group 2 mean	Effect size	Interpretation
42	44	0.50	Medium
42	46	1.0	Large
42	48	1.50	Large

Both groups were calculated with SD = 4.

Effect size for two teaching typing methods

A research team ran a study to two different methods of teaching typing skills. They found that method A resulted in students type an average of 42 words per minute (wpm) and method B students typed an average of 45 wpm. They found this difference to be statistically significant.

However, the effect size may be small because the standard deviation for the two samples can reasonably be expected to be large. As part of the data analysis, researchers cannot just report that the findings are statistically significant but should also include the effect size (is the effect of any practical significance). Based on both, they can draw conclusions about the study. At first glance, a study showing a difference of two teaching methods leading to 42 and 45 wpm results may not have a high practical significance. Researchers should conclude that although the study showed statistical significance, from a realistic view point, the two methods are equivalent.

The practical significance comes into play when for considerations of extending the study *t* larger sample sand populations. Perhaps the 42 wpm teaching method had a substantially lower cost, which would indicate it should be the method used. Or looking at the SD values, the method with the lowest SD could be recommended since it will provide the greatest number of student with relatively good typing skills. It is also important to look at the distribution because skewed distributions will strongly affect the practical significance of the study's conclusions. A variation of this example discussing skewness was shown on page 68.

Statistical Power of a Test

The effect size is calculated during the data analysis to show the practical strength of the results. Obviously, a researcher wants a study to have an acceptable effect size. Since the strength of the effect size is driven by the sample size, during the study planning phases, the proper sample size must be determined. This is done with a power analysis.

When designing a study, it is important to consider how many subjects to use. Too many subjects increase the time and expense. Too few subjects and the study risks making a type I error—rejecting a null hypothesis that is true. Statistical power is defined as the ability to find a statistically significant difference when the null hypothesis is really false. In other words, statistical power puts a value on the ability to find a difference when a real difference exists.

Power analysis is done early in a study. Based on the desired effect size, the power analysis gives the number of subjects required in each sample. A drawback is that the calculation requires the means and standard deviations for all subject groups. Often related research from the study's literature review can be used to make a reasonable estimate. There are many online statistical power calculators.

Power analysis calculations can also be used during a study. Collecting a subset of data will give an estimated mean and standard deviation that can be used to reevaluate the power analysis. Thus, you can collect data from 10 subjects, run the power analysis calculations, and find a total of 20 subjects per group are required.

Statistical power is the probability that the researcher will avoid making a type I error—and is selected by a researcher prior to the research project. If a statistical analysis has little statistical power, the researcher is likely to overlook or miss the desired outcome because the analysis did not have enough power to detect any significant differences. Thus, the power of a statistical test is a calculation of the test's probability of correctly rejecting the null hypothesis. The complement of the false negative rate (called β), which means the power of the test is $1 - \beta$.

Typically, a power analysis of about 80% is considered adequate. Power analysis depends on estimating the population mean, the experimental group mean, and the population standard deviation.

The power of a study is determined by four factors: the sample size, the alpha level, the effect size, and the standard deviation.

Sample size	In general, the larger the sample size, the higher statistical power in the analysis, but the larger sample size involves more time and expense for the study. There is also an issue that a large sample will often show statistical significant for very small differences in the sample means, which can be misleading. Thus, power analysis should be performed early in a study design to determine the minimum sample size that is required to detect an effect size. It can also be used to determine the minimum effect size that can be determined by a fixed sample size.
Alpha level	This is the *p*-value used to determine statistical significance, typically 0.05.
Effect size	Most sources say to use an effect size of 0.8. This gives an 80% chance of finding a statistically significant difference if it exists. The 80% chance may seem low, but the number of subjects required to obtain values closer to 100% increases rapidly.
Standard deviation	The standard deviation for the groups. Most power analysis calculators ask for both the mean and standard deviation of both the control and experimental groups. When running the power analysis before a study collects data, you can use previous studies found in the literature to provide a reasonable estimate.

Study designs that use subgroups need to do a power analysis to ensure that each cell in the design has enough subjects. A major factor affecting the reliability of a study is having cells with too few subjects for the power of the study. In Table 2.5, a study is looking at the effect of caffeine on test results versus time studying. Here the study is looking males and females separately and dividing them into groups with caffeine and no caffeine for two different study times. This gives eight separate cells that need to be populated. The statistical power test gives how many people should be in each cell.

Power analysis can also be run as part of the data analysis, typically when a result turns out to be nonsignificant. In this case, statistical power gives a means of verifying whether the nonsignificant result is due to really no relation in the sample or if the test lacked the of statistical power to determine a significant result. If the power analysis shows a sample size of 30 should have been used and

Table 2.5 Study design with multiple cells.

		1 h study time	3 h study time
Female	Caffeine		
	No caffeine		
Male	Caffeine		
	No caffeine		

If the power analysis calls for 15 subjects in each cell, a total of (15×8) 120 subjects would be required.

the study only had 10 subjects in each group, then the study lacked adequate power to make any claims about the hypothesis.

A study report should include the power analysis as a way of showing the number of study subjects was considered. Many funding agencies and grant proposals require a section showing the power analysis and justifying the proposed study sample size.

Example power estimate from a grant proposal

This is the power estimate that was part of a National Institute of Health (NIH) grant proposal. NIH requires this information be included in a proposal.

To estimate the number subjects needed, a power analysis assuming a normal distribution with unequal variances was performed.

Our goal is to have the multiple-choice questions written in a manner that prevents both ceiling and floor effects. This will be accomplished if 60% of the subjects answer each question correctly. Iterative rounds of prototyping questions will be used to verify this value. Since there are no comparable studies, estimates of the standard deviation was difficult. We used the values obtained by Cardinal and Siedler (1995) in their study of readability of healthcare material since the area of study is similar. The values from that study are SD = 12 for low groups and SD = 10 for high groups. They also saw a difference of 14% in total score between groups. Thus, the numbers we used for the power analysis were control mean = 53 SD = 12 and experimental group mean = 67 SD = 10. For a significance level of 0.05 and a power of 0.90, this gives a value of a minimum of 12 subjects in each cell of the test design.

To increase the power and to allow for the uncertainty in the standard deviation values, we intend to try to have 20 subjects in each cell. Using the same parameters as before, 20 subjects per cell give a significance level of 0.05 and a power of 0.98.

Practical Significance Versus Statistical Significance

Many people equate statistical significance to "significance" in the ordinary use of the word, meaning "useful" or "meaningful to me." The p-value is not, contrary to widely held beliefs (see Seth et al., 2009; Ziliak and McCloskey, 2008), an indicator of the practical importance of a finding.

In many studies, although the findings are statistically significant, the difference between the two groups is too small to be practically significant. For example:

- A travel company was considering upgrading the agents' displays with new and expensive displays. A cost analysis determined that to make the new displays worthwhile agents needed to enter reservations 10 seconds faster. A usability test of the old and new displays found the new ones were 3 s faster. The results were statistically significant, but not practically significant since the 3 s increase was slower than what would justify the cost of new displays (Lane, 2007).
- A drug study finds a new drug lowers blood pressure from 140/90 to 138/88. Although the decrease is statistically significant, the decrease is not clinically significant enough to justify using the drug.
- Newspaper reports of studies claim that a chemical causes a "significantly increased risk of cancer;" the study itself only reported they found a statistically significant difference. The total number cancer cases may have increased from a baseline of 10 per 10,000 to 13 per 10,000, possibly posing only a tiny absolute increase in risk.

Along with presenting a claim of being statistically significant, the research report needs to consider if that significance is enough to be practically significant for the readers. Of course, the meaning of practical significant differs between audiences, which must also be considered.

Statistical Independence

Two events are said to be independent if one event's occurrence does not influence the probability that the other event will or will not occur. Test of statistical significance assume that all of data are independent. The results of one test or trial do not influence future results. The study design needs to consider issues of statistical independence and the design should attempt to minimize their effect. For example, a study pretest and posttest cannot use the same questions, since the pretest cues the subject to look for the answers, violating the independence of the posttest.

Violating independence is different from hidden or known confounds in the data collection. A study of student problem solving ability in a game might be strongly influenced by the amount of video games they play—which you must allow for—but that does not affect the independence of the data collection. On the other hand, if previous subjects have talk about the test with upcoming subjects and give them tips, then the independence is compromised. The early trials affected the results on some of the later trials.

In the physical sciences, statistical independence is easy. Every time the ball in dropped in a physics experiment, the time it takes to fall in independent of the previous times. However, in social science study designs, achieving statistical independence can be difficult.

- Early questions on a survey questions can cause different answers to later questions.
- Order of tasks or learning curves in a test can influence later task since people learn how to do them.
- Interactions between subjects can lead to giving partial answers or similar answers.

Reusing people in a study can violate the independence of the test. This is why you need different random samples for each iteration of a study rather than reusing the same 20 people to repeat a study multiple times.

Statistical independence must also be considered in the interpretation of the data analysis results. The exploratory analysis could look at the data (answers, task times, etc.) plotted in the order they were obtained. Independent data should show a random scatter, while data without statistical independence may show a decrease in scatter over time.

Statistical independence in a problem solving task

A study looks people's problem solving ability by having them use a computer to manipulate a maze with obstacles. The time to complete the maze shows a graph like Figure 2.6.

The data analysis needs to consider if something influenced the statistical independence of the points that are showing the decreased time. Were those subjects talking to people who had completed the maze and getting tips: "when you get to the green door with the big padlock, just do this . . . .?"

Issues of violating statistical independence may not affect all of the collected data. The subjects that did not get advice had similar times to the early subjects, as expected.

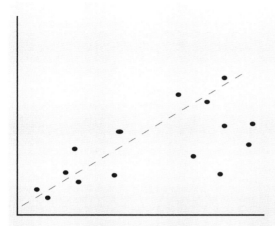

Figure 2.6 Time to solve the maze. Each point is one subject's time. Notice how the later data points seem to be diverging into two groups: one that fits the initial data and one that is consistently lower.

Even if the subjects did not get advice—compromising data independence—the data analysis also needs to consider if the subject pool contains two populations and that is the underlying cause of the different data points. The gamer and nongamer groups may have very different times.

Degrees of Freedom

Degrees of freedom are an index of the number of observations in a distribution that are free to vary from the mean. In a single series of numbers, such as would be applicable for a t-test, degrees of freedom equals $N-1$. Once $N-1$ values and the mean are known, the last value is fixed. Degrees of freedom are at statistic that gets reported in a study report, but with the use of computers is no longer a number that researches have to calculate.

Degrees of freedom differ for different arrangements of data and different statistical tests.

Example of degrees of freedom

Consider that a study as collected the task times for six people.
Number of observations: 6
Task time in seconds: 17, 26, 30, 35, 38, 45
Mean task time: 31.83333
Degrees of freedom: $N-1$ or $6-1=5$.
If we know the mean and know the value for any five values in the distribution, then the sixth number cannot vary.

Measures of Central Tendency

The common numbers associated with a data analysis—mean, median, standard deviation—are all measures of central tendency. In other words, descriptions of the center (most common) elements of the data.

Median

The median is the value that divides the distribution in half, such that half the values fall above the median and half fall below it. Computation of the median is relatively straightforward. The first step is to rank order the values from lowest to highest. If the data has an odd number of values, the median is the middle number. If there is an even number of values, the median is the halfway point between the center two numbers.

Nonparametric tests generally are based on the median, not the mean value of the data set.

Mode

The mode is the most frequently occurring values. In this sample, the mode would be 39 because a value of 39 occurs three times, more than any other value.

32	32	35	36	37	38	38	39	39	39	40	40	42	45

A distribution may have more than one mode if the two most frequent values occur the same number of times. Such distributions are called bimodal.

Mean

Mean (average), abbreviated with the Greek letter μ, is defined as the sum of numbers divided by number of data points.

Although the mean is easy to calculate, its use within a study requires you to consider the overall distribution. All four data distributions shown in Figure 2.7 have the same mean and could have the same number of data points. However, they are clearly all different distributions and require you to interpret the data differently.

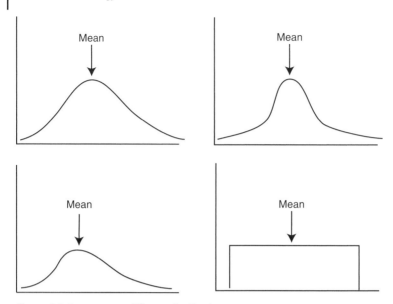

Figure 2.7 Same mean; different distribution.

Analysis means thinking in practical terms

Whenever statistics are used, it is important to think in terms of what they mean in practical terms and not blindly cite the numbers. A class assignment had students analyze sales records, which included data such as date of the sale, amount, numbers of items, etc. The sales showed a uniform distribution with the same number of sales and amount on each day of the month. Multiple reports contained a paragraph similar to this one:

> The average customer purchase date was around the 15th day of the month. The standard deviation for the purchase date was 7.66 days. This means that a sale could range from a week before the average purchase date to a week after. Also, it shows that a sale was likely to occur between the 7th of the month and 22nd of the month.

This type of statement results when numbers are blindly crunched without graphing or thinking about what the data means. The analysis averaged the day of the month (ordinal data) and noted the average sale occurred on the 15/16 of the month. If a store has a similar number of sales every day, then the average will always be the 14th–16th, unless something skews the data (big sale or closed for a snow storm). For that matter, if all of the sales occurred on the 1st and the 30th, the average would still be the 15th.

Rather than just reporting numbers, data analysis means to think through what the number means. This example shows why it is important to graph the data (which would reveal an even distribution), and not just crunch numbers to find the mean and standard deviation.

In the bigger picture, this sort of problem is common when (mis)interpreting ordinal data.

Variance

Variance measures the spread of the data points. A distribution with a high variance has data points that differ widely from one another. A distribution with a low variance has data points that cluster together around the mean. Formally, variance is defined as the average of the squared differences from the mean.

In modern practice, variance itself is typically not used much in data analysis, although computers can easily calculate it. However, standard deviation, which is used constantly, is defined as the square root of the variance.

There are different *t*-tests and ANOVA tests depending on whether the control and experimental groups have equal or unequal variance—is the data spread similar in the two groups. Statistical software contains multiple tests of variance, such as Leverne's test. This test examines the variance of the two samples and returns a result for whether or not they probably have the same variance. Unfortunately, small samples will almost always return a result of the data having the same variance.

Standard Deviation

Standard deviation (normally represented by the symbol σ or SD) is an indication of how much variation a set of numbers has from the mean. Formally, the SD of a distribution is the square root of its variance. A low SD indicates that the data points tend to be very close to the mean, whereas high SD indicates that the data points are spread out over a large range of values.

For a normal distribution, one SD from the mean always contains 68% of the data points (Figure 2.8).

The width of a normal curve depends on the standard deviation. Two different sets of data can have the same mean, but very different SD (Figure 2.9). But in both cases, 68% of the points are within one SD of the mean,

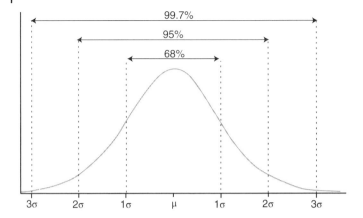

Figure 2.8 Standard deviation of a normal distribution. 68% of the data points will be within one SD. 95% of the data points are within two SD.

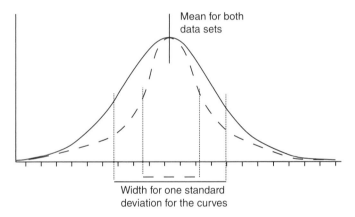

Figure 2.9 Different curve profiles for the same mean with different standard deviations. The one standard deviation points are marked by the vertical dotted lines.

but the points in the dashed line data set are clustered much closer to the mean since they have a smaller SD. However, for the wider distribution, those points can be much farther from the mean.

The SD value can be used to define outliers in the data. Unless a data set has a large number of points, data points more than 2 or 3 SD should be considered suspect. Table 2.6 lists the percent of data points found within different SD

Table 2.6 Percent of data points within different SD of the mean.

SD	Percent of data points within the range (%)	Expected occurrence of an value outside the range
1	68.27	1 in 3
2	95.45	1 in 22
3	99.73	1 in 370
4	99.99367	1 in 15,787
5	99.99995	1 in 1,744,278

values. Clearly, the chance of seeing a data point at 3 or more SD is remote unless there are large amount of data collected.

Many researchers have memorized that 68% of points equals one standard deviation and assume it is relevant to all situations. However, if the data does not fit a normal distribution, then the percentage values for standard deviation is not as meaningful. The claim that 68% of the points are within one SD or 95% within two SD is easy to show as false. For example, distributions with heavy tails [see page 65] would not have 68% of the data points inside of one standard deviation. Likewise if the data set was a bimodal distribution.

Percentile and Percentile Rank

Percentile and percentile rank look at the number of ordered data points above or below any specific data point. Percentiles are not concerned with the differences between values but with their position in the overall order. Many nonparametric tests are based on percentile and percentile rank calculations.

A percentile is the value of a variable below which a certain percent of observations fall. For example, the 20th percentile is the value below which 20% of the data points may be found.

The percentile rank is the percentage of scores in its frequency distribution that are the same or lower than it. Percentile ranks do not use an equal-interval scale; the difference between any two scores is not related to differences in value.

Consider the ordered set of 13 numbers (Table 2.7): 4, 6, 23, 45, 23, 42, 48, 52, 66, 68, 72, 99, and 133. The 50th percentile rank is number 48. But the distance between any two numbers is not the same.

Table 2.7 Percentile and percentile rank.

Data	Rank	Percentile (%)
4	1	100
6	2	92
23	3	85
23	4	77
42	5	69
45	6	62
48	7	54
52	8	46
66	9	38
68	10	31
72	11	23
99	12	15
133	13	8

Central Limit Theorem

A basic assumption of any research study and especially studies using statistical analysis is that the population mean of the dependent variable is the mean of many independent identically distributed variables. Or in simpler language, the real population mean is the mean that would be obtained from running the study many times and averaging each study's result. Statistical theory calls this the *central limit theorem* and also states that the means of those studies will form a normal distribution. Most of the common statistical tests use the central limit theorem to assume a normal distribution of the data.

The central limit theorem holds even if the distribution being studied is not a normal distribution. In other words, the distribution of the means of all of the studies will form a normal distribution, even if the data itself is highly skewed or otherwise nonnormal in each study.

If a study looks at task completion time for assembling a Lego model, the result would be a skewed distribution. A few might be faster, most people would take about the same time, and there would be a long tail of people who take longer (Figure 2.10). Remember that Figure 2.10 is a graph of the task completion time for a large number of individuals.

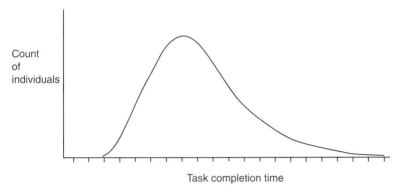

Figure 2.10 Time to complete a task normally is a skewed distribution. Very fast times are prohibited because of the nature of the task and some people take a long time.

The central limit theorem looks at the results running the study many times. A study measures the task completion time of 20 subjects and gets a result that is the average task time. Every time the study is run with a different set of 20 people (they must be different or at least all randomly selected from the population because of the requirement for independent events), the result will be a different average task time. If those mean times are plotted on a histogram with the mean task time as the x-axis and the count of the number of times that result occurred on the y-axis, a normal distribution will result (Figure 2.11). Remember this is a graph of all the means; if all of the individual task times were plotted, the graph would be the skewed distribution of Figure 2.10.

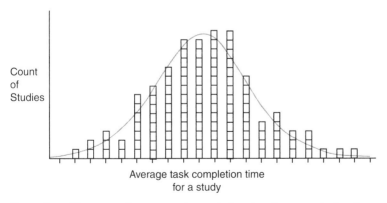

Figure 2.11 Histogram of average task completion time for many studies. Because of the central limit theorem, the distribution of these average times will be a normal distribution.

Law of Large Numbers

The law of large numbers states that the result of performing the same experiment a large number of times should be close to the expected value, and will tend to become closer as more trials are performed. In other words, the larger the data set, the closer the study mean will reflect the population mean. In terms of coin flips, a large number of flips with a fair coin will approach 50% heads and 50% tails. For a dice roll, all numbers will appear 1/6 of the time. Casinos make money because they can depend on the law of large numbers; some people may win big on a single roll, but in the long term, the odds will reflect the probability of the game, which favors the casino.

The law of large numbers assumes statistical independence (each coin flip is independent of previous ones). In a study design terms, it means that the data collected over a large number of events is better than the data collected over a small number of events.

Gambler's Fallacy

Although the law of large numbers is rather intuitive, it only applies to large samples. There is no principle governing small number of observations: there is no expectation they will converge or that any string of events in one direction will be balanced with a string in another direction.

People tend to misapply the law of large numbers to small sample sizes. They believe that small samples operate the same way and accurately represent the entire population. "They regard a sample randomly drawn from a population as highly representative, that is, similar to the population in all essential characteristics" (Tversky and Kahneman, 1971, p. 105). With a larger sample, the data will represent the population, but with a small sample, this assumption does not apply.

Making assumptions based on a small number of data points—essentially the opposite of the law of large numbers—is the gambler's fallacy. The gambler's fallacy is the belief that if deviations from expected behavior are observed in repeated independent trials, future deviations in the opposite direction are more likely. If a roulette wheel has three blacks in a row, a person may bet red because they think that after three consecutive black numbers that a red is more likely. Since each spin is independent, the three consecutive black numbers have no influence on the next spin. Likewise, if a fair coin has come up head three times in a row, the chance of a head on the fourth throw is still 50%

A fundamental assumption in statistics and probability is that events are independent of each other. People expect to find that any string of events that is statistically high or low is likely to be followed by events that move toward the mean. In the process, they ignore the independence of the events. Each toss of a coin has no connections with the previous tosses. People, on the other hand, often assume that past performance will predict the future. For example, a fair coin comes up heads eight times in a row. People see eight heads and, knowing that eight consecutive heads is a rare event, believe the odds of a tail on the next toss are greater than 50%. The probability of heads on the ninth toss is still 50%, since each toss is independent of the rest. With the gambler's fallacy, people use the eight heads in their prediction that the next toss must be tails. Note this is different from the chance of nine heads in a row; in this example, we have eight consecutive heads and are predicting the ninth toss.

A marketing survey that looked at the preference for brand X over brand Y and found that of the five people surveyed, four preferred brand X—although, only five people were surveyed, the probability of getting four out of five by chance is 3/8. For example, a coin flipped four times has a reasonable chance of four heads or four tails, but a coin tossed 100 times has a minimal chance of 100 heads.

During the exploratory analysis, it is easy to look at the data and see various patterns that are really the result of the small number of data points and a strong innate desire to see a pattern. People are mentally structured to see patterns; during the exploratory analysis, you must be careful not to overdo finding them. Or worse, to see spurious but desired patterns and start forcing the analysis to support those patterns.

Misunderstanding small numbers

Consider a survey with the following question:

Would 10 tosses or 100 tosses of a fair coin were more likely to have exactly 70% heads?

People tend to correctly choose the small sample. However, they have trouble transferring the abstract coin toss questions to real-world situations. When asked essentially the same question:

Would a large, urban hospital or a small, rural hospital be more likely to have 70% boys born on a particular day?

People answered that both hospitals were equally likely to have 70% boys born on that day. The large urban hospital will have lots of babies born on any particular day, which will drive its average closer to 50% boys. On the other hand, the rural hospital, like tossing a coin 10 times, will show large fluctuations in the boy/girl ratio on any one day, since it will have only a few births each day. But over the course of a year, both hospitals will deliver essentially equal percentages of boys and girls; as the numbers grow larger, they regress to the mean (Garfield and delMas, 1991).

Sports players do not have hot streaks

The chance of a hit in baseball is independent of the previous at-bats (Albright, 1993) and a basketball player's chance of making a shot is independent of his previous shots (Gilovich et al., 1985).

Albright (1993) looked at baseball hitting and found there are no hot or cold streaks but that the hitting exhibited a pattern expected by a random model. In other words, although it may seem like a batter is on a hot streak (is batting 0.600 for the week), when compared to his normal batting average, a seven game average of 0.600 is expected on occasion. He will also hit 0.200 for a week at some time during the season (a cold streak). Albright found the hitting pattern fit a bell curve and at times some sequence of consecutive hits will be outside of one standard deviation. But whether the current performance places the player in the standard deviation above or below normal occurs essentially the same. Likewise for making shots in basketball.

The hot streak appeared only by looking at a small sequence of the player's yearly or lifetime playing. The fans and players dispute this result, but the hot or cold streaks do not last longer than predicted by a statistical analysis.

Clustering Illusion

When the probabilities of repeated events are not known, outcomes may not be equally probable. Social science research does not know the expected mean for a study (which is the reason the study was performed), and a researcher can fall into a gambler's fallacy if a study returns data clustering around a specific point. Combined with a confirmation bias (where the researcher wants a particular result), a low number of data points can lead to misinterpreting the data.

The clustering illusion refers to the tendency to erroneously perceive small samples from random distributions as having significant clusters (Figure 2.12). From a probability theory perspective, a small sample of random or semi-random data has a high probability of showing strings of a single event. One reason to use inferential statistics to analyze data is because it tends to not be affected by the random clusters. Of course, the researcher must not decide the clusters are real and ignore the analysis results; instead, any strong disagreements must be carefully evaluated.

When looking strings of numbers or other events, people see clusters of events (good or bad) and assume they cannot be random. People are hardwired to see patterns and have a hard time recognizing truly random information. A string of numbers or a scatter plot may be random, but people will see patterns

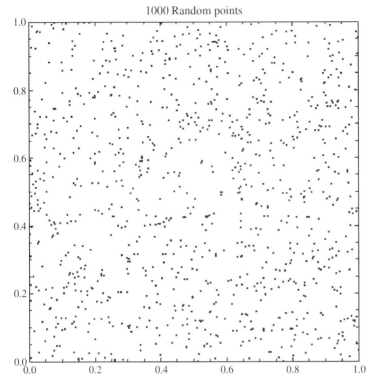

Figure 2.12 Cluster illusion. This image has 1000 points randomly distributed, but most people see clusters of data points and blank areas with no data. In a data analysis, they risk assuming the clusters are meaningful and interpreting the data incorrectly. (Generated from http://en.wikipedia.org/wiki/Clustering_illusion.)

in it. If asked if a sequence is random (e.g., TTHTTHHHHTTHTTHTH), people will often confuse random sequences and nonrandom sequences. Part of the problem is that random sequences will have strings of events that people see as nonrandom. In the example just shown, the four Hs might make some people declare it as nonrandom.

In data analysis, seeing strings of events in the data can make a researcher misjudge the randomness of the data and try to attach extra significance to those strings.

Also, people cannot create a random set of numbers. When asked to generate lottery number sequences, the sequences that people created were not random, even when they were told to create a random sequence (Boland and Pawitan, 1999). If people are asked to create a random string of 50 heads/tails, the resulting string will usually fail a random number test.

References

Albright, S. (1993) A statistical analysis of hitting streaks in baseball. *Journal of the American Statistical Association*, **88** (424), 1175–1183.

Boland, P.J. and Pawitan, Y. (1999) Trying to be random in selecting numbers for Lotto. *Journal of Statistics Education*, **7** (3). Available at http://www.amstat.org/publications/jse/secure/v7n3/boland.cfm.

Cardinal, B. and Siedler, T. (1995) Readability and comprehensibility of the "Exercise Lite" brochure. *Perceptual and Motor Skills*, **80**, 399–402.

Coe, R. (2002) It's the effect size, stupid: what effect size is and why it is important. *Paper presented at the Annual Conference of the British Educational Research Association*, University of Exeter, England, 12–14 Sept. 2002. Available at http://www.leeds.ac.uk/educol/documents/00002182.htm (accessed May 29, 2016).

Cohen, J. (1994) The earth is round ($p < .05$). *American Psychologist*, **49** (12), 997–1003.

Garfield, J.G. and delMas, R.C. (1991) Students' conceptions of probability. In: D. Vere-Jones (Ed.) *Proceedings of the Third International Conference on Teaching Statistics*, Voorburg, The Netherlands: International Statistical Institute, vol. 1, pp. 338–339.

Gilovich, T., Vallone, R., and Tversky, A. (1985) The hot hand in basketball: On the misperception of random sequences. *Cognitive Psychology*, **17**, 295–314.

Jones, L. V. and Tukey, J. W. (2000) A sensible formulation of the significance test. *Psychological Methods*, **5**, 411–414.

Kline, R.B. (2004) *Beyond Significance Testing: Reforming Data Analysis Methods in Behavioral Research*. Washington, DC: American Psychological Association.

Lane, D. (2007) Statistical and practical significance. In *HyperStat Online Statistics Textbook*. Available at http://davidmlane.com/hyperstat/B35955.html (accessed May 12, 2009).

Nuzzo, R. (2015) Scientists perturbed by loss of stat tools to sift research fudge from fact. Available at http://www.scientificamerican.com/article/scientists-perturbed-by-loss-of-stat-tools-to-sift-research-fudge-from-fact/.

Seth, A., Carlson, K.D., Hatfield, D.E., and Lan, H.W. (2009) So what? Beyond statistical significance to substantive significance in strategy research. In: D. Bergh and D. Ketchen (Eds.) *Research Methodology in Strategy and Management*, New York: Elsevier, vol. 5, pp. 3–28.

Siegfried, T. (2010) Odds are, it's wrong: science fails to face the shortcomings of statistics. *Science News*, **177** (7), 26.

Trafimow, D. and Marks, M. (2015) Editorial. *Basic and Applied Social Psychology*, **37**, 1–2.

Tversky, A. and Kahneman, D. (1971) The belief in the law of small numbers. *Psychological Bulletin*, **76**, 105–110.

Ziliak, S.T. and McCloskey, D.N. (2008) *The Cult of Statistical Significance: How the Standard Error Costs Us Jobs, Justice, and Lives*. Ann Arbor, MI: University of Michigan Press.

3

Analysis Issues and Potential Pitfalls

A researcher must take the results of the various tests and evaluate them with respect to the data and the research context. The deeper the researcher looks, the better the data analysis. This chapter looks at some of issues that must be considered as part of that deeper look.

Some people want to run a statistical test and take the *p*-value as the answer of significance or not, with no further analysis, and write a report. Others want to avoid doing any inferential statistics and simply examine the data to reach an answer. Both analysis methods are insufficient and lead to unsatisfactory data analysis. Data analysis is a complex problem that does lend itself to a simple answer; both the examples above tried to oversimplify the process.

Effects of Variables

Main Effect

In the formal definition of the design of experiments and data analysis, a *main effect* is the effect of an independent variable on a dependent variable averaged across the levels of any other independent variables. The dependent variable changes based on both the independent variables and the interactions between those independent variables.

In writing up research, the term main effect is used to explain how the major variable being studied influenced the dependent variables and to distinguish it from other interaction effects. If there are not interacting factors, "main effect" typically is not used, since it then becomes equivalent to "independent variable."

In more practical terms, many studies have multiple factors interacting that determine the final value of the dependent variable. A laboratory study may be able to control some of these, but a study done in a more natural setting must include them in the analysis. For example, if we wanted to study strenuous exercise on answering a timed-essay question, we could have people perform a

Introduction to Quantitative Data Analysis in the Behavioral and Social Sciences,
First Edition. Michael J. Albers.
© 2017 John Wiley & Sons, Inc. Published 2017 by John Wiley & Sons, Inc.
Companion website: www.wiley.com/go/albers/quantitativedataanalysis

strenuous exercise for 2, 4, 6, 8, and 10 min. Using an ANOVA, the main effect would be exercise time. However, there are other factors to consider and that may or may not interact with the other data that was collected (fitness level, knowledge of the essay subject, typing speed, room noise, etc.). Each of these can influence how the essay question is answered. It is possible that some of these factors may be more relevant than the study's main effect—we might find knowledge of the essay subject and typing ability are the two main factors in the quality of the answer and level of strenuous exercise does not matter.

Working out the interacting effects of multiple variables requires a careful analysis and may require looking at the data in multiple ways.

Data analysis definitions must be resolved early

A quantitative study that uses archival data has to carefully define how the data were be analyzed. Consider this example on credit card debt.

"What does the average Joe owe on his credit cards? That would be $3,600. Or $7,743. Or $5,234. Or $1,098" (Williams, 2015).

The data analysis issue is that all of these numbers are correct, but they are all calculated differently. Is that all credit cards total or per card balance?

- The Federal Reserve uses debt of people with a credit card and a Social Security number, so noncitizens are not counted.
- TransUnion does not use cards that have gone unused for a long time in per card balance. Two cards with $5000 balance and three unused cards average to $1000 balance per card.
- Only cards with balances (not those paid off each month) are used.

Each of these is a valid way to calculate mean and median of the data set, but each return very different numbers. How those calculations were made must be described in the study report. And, most importantly, the study must only make claims that reflect the limitations of the calculations.

Data Interactions

Data interactions mean the effect of one independent variable may depend on the level of the other independent variable.

Many statistical tests, in the case of this book primarily the ANOVA, have options for one-way, two-way, and perhaps three-way (or more) interactions.

A study looks at the quality of writing during a timed writing. The independent variables are time (give people two different time periods to write) and typing ability (they were given a typing test and grouped into a low, medium, and high group). There are two variables, so there are two main effects: more

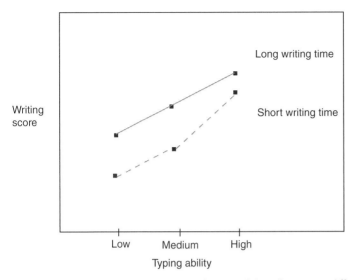

Figure 3.1 Interaction of writing time and typing ability. The varying differences in the two lines are an indication there may be an interaction.

time could improve the quality and better typist could write better. But with an interaction these two factors may not simply add together. The time and typing ability may interact (affect each other) so that together they create a third term that influences the result of the final writing quality. Thus, the final quality is determined by three terms: writing time, typing ability, and writing time × typing ability. Graphically, the data may look like Figure 3.1.

An interaction is a term in a statistical model in which the effect of two, or more, variables is not simply additive. If Figure 3.1 had parallel lines, then there would be no reason to suspect a data interaction. In a two-way interaction, the relationship between the independent variable and the dependent variable is affected by a third variable (in the example, the third variable is typing ability). The third variable can be either an independent variable or a fixed factor such as subject age. Higher level interactions (three-way, four-way, etc.) are similar, but with more variables affecting the dependent variable. If the study had also included amount of sleep as a variable, there may also be an interaction of sleep with typing and writing ability. If a study's data analysis seems to involve three-way or more interactions, consult a statistician. Properly handing the variables in statistical software packages is not straightforward.

The simplest, a one-way interaction, tends to imply causality and is called the main effect. A two-way or more interaction loses simple causality because of how the various independent variables affect each other. Note: in the regression test, the product term (independent variable × third variable) will be significant if there is an interaction between them.

Example of interactions between font size and layout

If we were examining the effect of two variables, font size and 1 or 2 column layout, on text comprehension, there would be four cells in the design matrix. The study design uses all combinations, with an equal number of subjects in each cell.

	One column	Two column
Small font		
Large font		

The data analysis could describe two main effects:

- Any difference in comprehension scores between font sizes. Perhaps people reading a smaller font size have a lower comprehension.
- Any difference in comprehension scores between column layouts. Perhaps people reading two-column text have a lower comprehension.

The presence of an interaction effect implies that the effect of font size on comprehension varies as a function of columns. In this example, it might mean larger fonts with two columns had a lower/higher comprehension than would be predicted by just looking at either font size or layout alone. The combination of font size and column layout *interacted* to create a third combined effect on the comprehension of the information.

Knowing that there are interactions has important implications for interpreting statistical models. If two variables interact, the relationship between each of the interacting variables and a third "dependent variable" depends on three values: the value of the two primary variables and product of the two primary variables. In practice, this makes it more difficult to predict the consequences of changing the value of a variable. When there are several variables, there can be several interactions. The interactions may show non-linear relationships or may be difficult to measure in practical situations.

Interactions in a study of Fitts' law

Fitts' law states that it takes longer to click on a smaller target. Experiments on Fitts's law have an independent variable of the size of the target and the dependent variable of the time required to move the mouse and click on the target.

A basic statistical test showing this simple relationship would be a one-way interaction: time versus size. If the age of the subjects was included, there could be a two-way interaction. But a two-way interaction is more than a finding such as "older people are slower." A study could find that older people on average take 1.4 times as long to click on a target. If that value applies to all target sizes, then it is still a one-way interaction. An age-target size interaction could mean that older people take 1.6 time as long to click on small targets, but only 1.2 times as long to click on large targets.

For a two-way interaction, the regression equation would show three factors: target size, age, and target size × age. The two factors interact in some manner. In this study, it might mean for a large target, age has minimal effect, but as the target gets smaller, then age has an increasing effect on time to click. Thus, rather than older people always needing 1.4 times as long to click, for a large target they might need 1.2 times as long and for a tiny target 3.4 times as long.

See Bakaev (2008) and Bohan and Chaparro (1998) for some actual studies that show age interactions with Fitts' law.

Outliers in the Dataset

An outlier is formally defined as "an observation that lies an abnormal distance from other values in a random sample from a population"[1]. For a detailed discussion of outliers, see Barnett and Lewis (1994).

Outliers are values that do not appear to fit within a data set because they are too low or too high compared to the rest of the data. How an outlier is handled depends on the potential cause. General data analysis guidance is to not ignore or discard outliers unless there is a clear and justifiable reason to do so. The study report should discuss how and why outliers were removed.

Outliers and how to handle them are discussed in more detail in "Handling outliers in the data" on page 177.

Relationships Between Variables

The normal goal of running a study is to uncover any relationships between the variables. The data analysis focuses on determining how the independent variables affect the dependent variables. A study can find three different relationships: no relationship, a cause–effect relationship, or a correlation relationship.

---

1 NIST/SEMATECH e-Handbook of Statistical Methods, http://www.itl.nist.gov/div898/handbook/prc/section1/prc16.htm

In data analysis, a researcher tries to resolve three different relationship issues:

- The existence of a relationship (statistical significance)
- The degree of the relationship (correlation)
- The type of relationship (the direction or mathematical equation, such as positive linear)

Statistical Significance—There is a Relationship

A test that gives a result of statistical significance indicates there is a relationship between the independent and dependent variables in the study. A change to the independent variable meant the study found a change in the dependent variable.

Not Statistical Significance—There Is No Relationship (Not Really)

No relationship (not statistically significance) may appear at first glance as a trivial finding and one in which the experiment failed. But experiments are not failures if they fail to find a relationship. Anytime the null hypothesis is not refuted (p-value > 0.05), the conclusion is that the study did not have the power to determine if there was a relationship between the variables. Note that this is very different from a claim of no relationship. A study cannot positively claim there is no relationship; it can only claim it was unable to find one. A new study with more subjects, more sensitive measurements, or a longer time period may find there is a relationship. A positive result can show a relationship, but a negative result only means this study did not find a relationship and does not say that one does not exist.

For example, if a study looked at a random selection of ten 55-year-old smokers and ten 55-year-old nonsmokers, because of the small sample size the study would probably not show any statistical significance for adverse health effects from smoking. The study can only say it was unable to find a relationship, not that one does not exist. Likewise, a study on a new writing method that looks at two class sections may not find any significance or relationship because of lack of sample size.

Cause and Effect

A basic purpose of many quantitative studies is to uncover cause–effect relationships. Unfortunately, within the social sciences cause–effect relationships are difficult to clearly find. When working with people, an untold number of other factors (typically uncontrollable and/or unknown) come into play that makes any strong claim of cause–effect questionable. However, it is still possible to get indications of how some change in the independent variable caused a change in the dependent variable.

A statistical analysis of properly collected experimental data provides a basis to claim that one thing causes another. They try to explicitly connect the cause (when A happens) with the effect (then B happens). It may be deterministic (gravity always causes objects to fall), but in most cases within the social sciences, the best that can be done is a statistical relationship (If A happens, then there is a 70% chance that B happens). Because these are statistical relationships, they cannot be applied to individuals, but they do describe a population.

- Having high blood pressure causes an increased risk of stroke.
- Studying before a test increases the score.
- Alcohol slows reaction times.

When a study's report includes cause and effect relationships, part of communicating the study's results involves the reader identifying the chain of events that give rise to that cause and effect. As such, this chain of events must be clearly laid out in the discussion section. This leads to the generalizable aspects of the study. Of course, since studies are limited in scope, it can difficult to resolve an issue such as sport cars accidents when the data only has car color. Part of the later stages of data analysis and connecting it to the research situation is figuring out which factors have not been completely covered and determining the future research paths required to answer those questions.

People inherently think in terms of linear relationships, but many data relationships are not linear—doubling the cause value does not imply a doubling of the effect value. The interpretation problem is that people tend to mentally and visually use linear extrapolation and are very poor at nonlinear extrapolation (Wickens and Hollands, 2000). Figure 3.2 shows some examples of nonlinear cause–effect relationships.

Correlation

Correlation shows a relationship between two variables; in other words, a change in one variable is reflected in the other.

Not finding relationships is not indicative of a study failure. In most social science studies, there are too many factors that cannot be controlled to allow for a clear relationship. Instead, the researcher can only show a strong correlation. Of course, the study report must be clear that the findings are a correlation and not a cause–effect.

Correlation relationship simply says that two things change in a synchronized manner and measures the strength of the relationship between the factors. How closely any changes in the independent variable are reflected in the dependent

Linear relationships

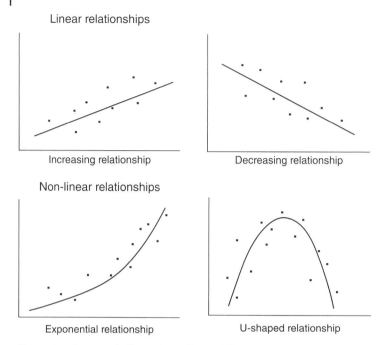

Increasing relationship   Decreasing relationship

Non-linear relationships

Exponential relationship   U-shaped relationship

Figure 3.2 Cause and effect relationships. All four graphs show a strong cause–effect relationship; however, all cause–effect graphs are not straight lines.

variable is the strength of the correlation. If a doubling of the independent variable causes the dependent variable to also double, the correlation is very strong. On the other hand, if the dependent variable increases by 5–20% across several subjects, then the correlation is weak.

However, **correlation does not mean causation**. In a correlation relationship, A and B are both present, but do not cause the other to occur; they may both be caused by another factor C. Task times for a web page with black text and a white background may be faster than task times for a web page with black text and a dark blue background, but the background color is not the cause of the slower times. Rather, they are only correlated. The cause is the vision impairment caused by the low contrast between the text and background. Another trivial example is that high school education achievement is related to the number of bathrooms in the student's home. Clearly, having more bathrooms will not make a better student. However, both are strongly correlated with a third variable, parent's economic status, which is a cause. Higher income people live in bigger houses with more bathrooms and students whose parents have a higher economic status tend to do better in school.

In the student achievement versus number of bathrooms example it is relatively easy to see that there must be a third variable that is the real cause.

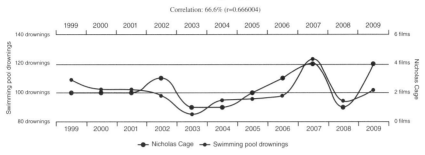

Figure 3.3 The graph shows a strong correlation of two factors that clearly have no connection with each other (Vigen, 2015).

But in many studies, determining that third variable or even realizing that it exists can be difficult. It would be easy to perform a study that lists the effect of different web page background colors on reading speed without considering the real cause driving the effect. In these cases, unknown variables within a study might mislead a researcher to thinking the study has found a cause–effect when it is really a correlation.

Figure 3.3 shows a spurious correlation between the pool drowning and Nicolas Cage movies. These factors clearly have no connection with each other and there is no reason to even suspect they are correlated. (Examples such as this are easy to find with a web search.) Look at enough variables, and correlations will be found among them. The question a researcher must answer is if they are valid correlations.

As with cause–effect relationships, correlations can be either linear or nonlinear. Many of the correlation tests check for linear correlation (a straight line relationship of both variables increasing/decreasing by the same relative amount). However, many variables display nonlinear correlation. Figure 3.4 shows a very strong correlation between the two variables, but the Pearson correlation coefficient for the nonlinear relationships may be zero or close to zero, implying no relationship. Thus, in correlation studies, it is essential to examine graphical plots of the variables and to examine the residuals rather than simply crunching numbers.

News reports often confuse correlation and causation. They may report that researchers have found a cause for a disease, when the research study itself only says it found a correlation. For example, news headlines proclaim that eating certain foods or specific behaviors cause a heart attack. The study they draw from reports a correlation that people who eat a lot of a certain food have a high

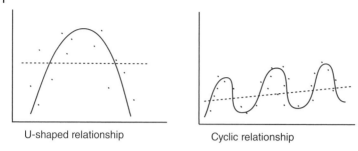

U-shaped relationship   Cyclic relationship

Figure 3.4 Scatterplots of linear and nonlinear correlations. The dashed line is the linear relationship calculated for the data. Nonlinear relationships may show a "no correlation" result when calculated with a Pearson correlation test, which tests for a linear correlation.

incident of heart attacks, but the news media switches a correlation to a cause–effect finding. The new reports say a study claims eating a certain food will cause a heart attack and then imply that not eating those foods will prevent a heart attack. The formal study report must not fall into the trap of confusing correlation and causation, which can be difficult if the researcher set out to find a cause–effect relationship.

Correlation of women executives and corporate financial results

Comparing the annual financial results for US corporations against the number of women they have in senior position shows a strong correlation: more women executives relates to better results.

> Organizations that include a high percentage of women in senior positions show better financial results. Companies in the top quartile for women in the executive committee from 2007–2009 had 41% greater return on equity, and 56% greater earnings before interest and taxes than companies with no women in the executive committee, for companies within the same industrial sector. Financial results for companies with at least three women serving on the board of directors are better: in 2007, return on equity was 16.7%, as opposed to an average 11.5%; return on sales was 16.8%, as opposed to an average 11.5%; return on invested capital was 10%, as opposed to an average 6.2%. (Nelson, 2014)

However, these data show a correlation, not cause. Simply hiring or promoting women to senior positions will probably not result in better financial results. Instead, the underlying factors—probably strongly dependent on the personalities of women who get promoted to senior positions and overall corporate culture—would be the underlying causes.

Statistical significant and correlation

At study looked at how spicy men liked their food and their testosterone level. Interestingly, they found a statistically significant relationship between "spice-love and testosterone level" (Bègue et al., 2015).

However, the study reported a correlation factor of 0.29, which is very low. It also found no causal factors. Thus, although the study could report a statistically significant finding, the practical significance was very low.

Confusing Linear and Nonlinear Relationships

Figure 3.5 has enough points to clearly show the relationship curves. Unfortunately, the small number of data points in most social science studies makes it harder to see the clear patterns evident in these scatterplots. Figure 3.5 contains two relationships with a smaller number of points that make it difficult to discern the actually relationship. Linear and exponential relationships are especially difficult with only a few points; the human mind wants to always see a linear relationship. Examining the residuals of a curve to the data points between different relationships can help to identify the one that matches the data best.

People mentally assume linear relationships and during the exploratory analysis phase will see linear relationships in scatterplots that are actually showing nonlinear relationships, especially when the total number of data points is small. The problem with assuming a linear relationship is that the difference between reality and the assumed relationship line can rapidly deviate and result in very poor predictions.

A second problem can arise when a study lacks sufficient breadth to determine the real relationship. If the data is collected over a small range, the difference between a linear and nonlinear relationship may be smaller than

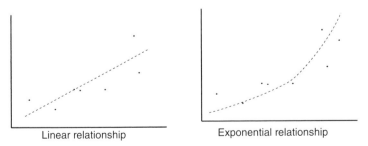
Linear relationship   Exponential relationship

Figure 3.5 Too few data points to visually see the relationships. Both graphs have the same data points, but different relationship lines drawn in.

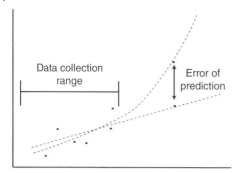

Figure 3.6 Problems of assuming linear relationships when the real relationship is nonlinear. Although a linear relationship may be close enough over parts of the graph, in the bigger picture it can deviate substantially.

the study's power. In this case, the study will conclude that the relationship is linear. If the conclusions are applied within the same limited scope as the study (Figure 3.6), the assumption can hold. But if applied outside of that range, there can be significant deviation. A study that looked at values between 10 and 30 can be used to make predictions within that range, but using a regression model to predict what happens when the value is 60 may or may not reflect reality.

A Single Contradictory Example Does Not Invalidate a Statistical Relationship

All too common, after reading about a research study people make comments like "they may have found that in their study, but I know two people at work who do not do it that way. Thus, this study does not apply to my work." or "this study is totally wrong."

The fallacy here is that they are trying to refute a statistical finding with a specific example. That is logically impossible. For example, high school education achievement is strongly correlated with parent's income; children of higher income parents do better in school. Pointing out that John, whose parents make $400,000, is a bad student and constantly in trouble or that Susan, whose mom makes $15,000, just got accepted to Yale does not invalidate the findings. If we take 1000 students whose parents earn $400,000 and $15,000, we would find the high-income students do better. Statistics only apply to a group; they say nothing about John or Susan as individuals within that group.

Deterministic cause–effect can be disproven with one counterexample, such as a ball not falling when dropped. Statistical cause–effect relationships—the normal result in social science studies—cannot be refuted by one counterexample; they deal with the group as a whole, not individual members. Many people try to reject a statistical cause–effect with anecdotic evidence about one situation they have experienced. But one specific example does not reject a statistical cause–effect relationship. Insurance companies know that people who drive red cars have more accidents. Comments about knowing a person who drives a red car and who has never had an accident does not refute the fact that red cars have more accidents. (Actually, it is not red cars that are accident prone. Red is a common sports car color and sports cars have more accidents. The cause–effect is sport cars accidents; the correlation is red cars accidents.) Likewise, low-income people who get accepted to Harvard do not disprove the cause–effect relationship of low income and low performance in high school; a statistical relationship does not say anything about a particular student, just the group as a whole. If 70% of the people are expected to successfully complete a task, the 70% probability says nothing about whether a specific individual will be in the 70% or the 30%.

A statistical relationship says that some percentage of a group (not every member) will fit. If I set up tall rectangular blocks and then hit the table, some of the blocks will fall over. If I do this many times, I may find that on average 80% of the blocks fall over. So I can confidently predict that when I hit the table again, that about 80% of the blocks will fall over.

However—and this is where the faulty logic occurs for most people—

- I cannot make any prediction about a specific block on any single hit. I paint one block purple and the others green. Over many trials, the purple block will fall down 80% of the time. On any specific trial, I cannot make a prediction about whether the block will fall or not. Sometimes, it will be in the 80% that fall and other times it will be in the 20% that do not fall.
- I cannot predict exactly how many blocks will fall over. Rarely will exactly 80% of the blocks fall over. There is no way to predict the exact actual number of fallen blocks. However, if I repeatedly hit the table, the percentage of fallen blocks will graph in a normal distribution with a mean of 80%. So, if only 74 or 84% percent of the blocks fall over in one trial, my average prediction is still valid.

If we go back to the previous example of John and Susan, they were both in the smaller percentage of students with academic performance that does not match their parent's economic status. But just like the blocks, with a statistical relationship, you cannot point at one element and disprove the relationship.

Let us look in more detail at the block problem. It is really the same problem as almost any study that looks at "when X happens, will Y occur? And how often will Y occur?" Every time the table is shoved, some of the blocks fall over. A study with 10 trials is not enough to give good probability of falling based on block location—a power analysis or effect size calculation will be required to get an estimate of how many trials to run. But if we do a lot of trials, the data analysis might reveal that chance of a block falling depends on its position (Figure 3.7). The initial data analysis looked at the blocks as a whole and counted how many fell. We can also do a deeper analysis based on individual blocks to determine if position matters.

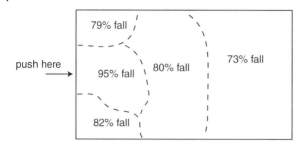

Figure 3.7 A blocks chance of falling by table position. With enough trials, a data analysis might result in a percentage map like this.

Being able to analyze the blocks falling by position is an issue of proper data collection, when needs to be defined early in a study. If just the total number left standing is recorded, then the analysis can never create a map as found in Figure 3.7 because it lacks the detailed data. In the block study, collecting individual block data is easy, but in many studies even the detailed collection, because of time or expense, must be at a higher level. Perhaps the table is divided into six areas and the number of blocks that fall in each area is recorded, rather than individual blocks.

References

Bakaev, M. (2008) Fitts' law for older adults: considering a factor of age. In *Proceedings of the VIII Brazilian Symposium on Human Factors in Computing Systems*, New York: ACM, pp. 260–263.

Barnett, V. and Lewis, T. (1994) *Outliers in Statistical Data, 3rd edn*, New York: Wiley.

Bègue, L., Bricout, V., Boudesseul, J., Shankland, R., and Duke, A. (2015) Some like it hot: testosterone predicts laboratory eating behavior of spicy food. *Physiology & Behavior*, **139**, 375–377.

Bohan, M. and Chaparro, A. (1998) Age-related differences in performance using a mouse and trackball. In *Proceedings of the Human Factors and Ergonomics Society 42nd Annual Meeting*, Santa Monica, CA: HFES. 152–155.

Nelson, B. (2014) The data on diversity. *Communications of the ACM*, **57** (11), 86–95.

Vigen, T. (2015) Spurious correlations. Retrieved from http://tylervigen.com/spurious-correlations.

Wickens, C. and Hollands, J. (2000) *Engineering Psychology and Human Performance*. Upper Saddle River, New Jersey: Prentice Hall.

Williams, F. (2015) Average credit card debt statistics. Available at http://www.creditcards.com/credit-card-news/average-credit_card_debt-1276.php (accessed Sept. 25, 21015).

4

Graphically Representing Data

In an ideal world, data would always plot on the desired graph line. But even in that ideal situation, the shape of the graph line varies. Many random events plot into a normal distribution (bell-shaped curve), other plot as straight line, an exponential line, or a u-shaped line.

The shape of the data distribution and the scatter around the ideal (or best fit) line determine which type of statistical analysis will be valid for the data. Thus, it is important to graph the data as part of the exploratory data analysis to understand how the various data elements relate to each other.

Data Distributions

Data distributions can come in many shapes (Figure 4.1). Early exploratory analysis of data should graph the results to give a feel for how or if the data are related.

People cannot look at tables of numbers and get a good feel for how they are varying across the study. Thus, early in a data analysis, it is important to graph the data and to get a feel for how the variables are interacting. The graph points to what tests are relevant and gives a feeling for what results to expect. They can also reveal some unexpected relationships to explore in the later cycles of the data analysis.

Many discussions of analyzing research data focus on the normal distribution and it is often the only one mentioned. The typical parametric statistical tests (i.e., t-test and ANOVA) assume a normal distribution. If other distributions are mentioned, they tend to be discussed in mathematical terms and not in terms of applicability to study analysis. Granted, a large percentage of research data is normally distributed, but there are other distributions. And some specific data types cannot be normally distributed—for example, gender of test subjects, survey questions where any of five answers are equally likely, or data that has

Introduction to Quantitative Data Analysis in the Behavioral and Social Sciences,
First Edition. Michael J. Albers.

Types of data distributions

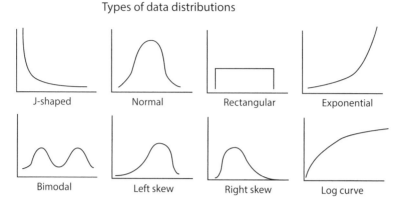

Figure 4.1 Types of data distributions.

yes/no values. Proper analysis must consider how the data should be distributed since the distribution drives the selection of the proper analysis method.

Bell Curves

A bell curve has a smooth rise and looks the same on both sides.

Normal Distribution Curve

A normal distribution is the familiar bell curve that forms the basic assumption of most parametric statistical tests (Figure 4.2). It describes a large range of naturally occurring phenomenon from people's heights to standardized test scores (e.g., SAT or ACT). The normal distributions differ from each other by their midpoint (the data set's mean) and the variability of scores around the midpoint (the data set's standard deviation).

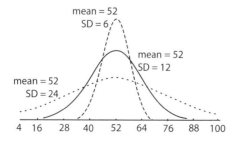

Figure 4.2 Normal distribution curves. All the curves displayed here fit a normal distribution. They all have the same mean (52) but appear different because of the different values of the standard deviations.

All the members of the family of normal distribution curves share a number of properties.

Shape
: Although the curves with different parameters look different, if the *x*-axis of the two curves is scaled properly, they look alike. Although this is an interesting mathematical property, it has no direct application to data analysis.

Symmetry
: A normal distribution has bilateral symmetry; the two sides are exactly alike.

Tails approaching but never touching the *x*-axis
: The tails have a mathematical limit of the *x*-axis, but they never actually touch it. The implication of this property is that no matter how the curve is drawn along the *x*-axis, in either the positive or the negative direction, there will still be some area under any normal curve. However, distances beyond 4–6 standard deviations from the mean are rarely relevant to social science research data analysis because obtaining events this far from the mean requires very large data sets. For example, 3 standard deviations means there is a 0.3% (three times out of 1000 trials) of an event occurring.

Every Bell Curve is not Normal

Every curve that has a bell shape is not a normal distribution; in other words, a bell shape does not imply normality. Normal curves are mathematically defined by an equation, which happily is never used during the analysis.

Figure 4.3 shows some curves that look normal, but are not; they do not fit the mathematical definition.

The fundamental problem from a data analysis perspective is that the change in shape can affect basic statistical issues, such as standard deviation. Data that display heavy tails will have more data points away from the mean than expected for a normal distribution.

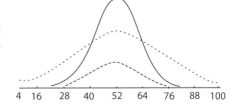

Figure 4.3 Bell-shaped, but nonnormal curves. At first glance, these distributions look like a normal distribution, but they fail to fit the equation defining a normal curve.

4 16   28   40   52   64   76   88   100

If a data set is not normally distributed, then the standard deviation is not defined and—although every statistics program will calculate it—reporting it can be misleading. Each statistical test has a different level of robustness for how well it handles the departure from normality. Luckily, most common statistical tests handle some departure from normal data quite well.

Testing Data for Normality

Parametric statistical tests assume that the data follow a normal distribution. Thus, it is important to check the data early in the analysis to verify it. Graphing the data gives a visual check, but as we saw in the previous section, the visual appearance alone is not enough to know if a distribution is normally distributed. There are some rigorous mathematical tests for normality. All the major statistical analysis software supports at least one of these tests.

The easiest initial check for a normal distribution is to graph the data and decide if it is close enough to a normal distribution to use parametric tests or if nonparametric tests need to be considered in the analysis. The visual inspection of the data easily shows if it is not normal (skewed, bi-modal, heavy tails, etc.). If the data look like a normal distribution, then a formal verification test can be run.

Unfortunately, in many studies, the low number of data points prevents getting a clear picture from the graph. Figure 4.4 shows a graph of 10 points that were sampled from a normal distribution; they are too spread out to give any visual support either for or against.

There are three main formal tests to verify that a data set fits a normal curve. They test how closely the data set corresponds to the expected data set if the data were sampled from a normal distribution. The goal for the tests is to have the null hypothesis supported; in other words, a large p-value is good since it means the data fit a normal distribution. However, tests for normality must be used with caution as they suffer from the following problems:

- They require a larger data set than is common in social science data.
- Small data sets will almost always pass the test. With small sample sizes of 10 or fewer, it is unlikely a normality test will detect nonnormality.
- In large data sets—with over 1000 data points—very small differences may show a statistically significant deviation from normality, but the deviation is not enough to affect the results of a t-test or ANOVA.

Because of these issues, many statistical sources recommend against routinely testing all data for normality. Instead, clearly justify why the test is being performed and how the results (positive or negative) will affect the rest of the analysis.

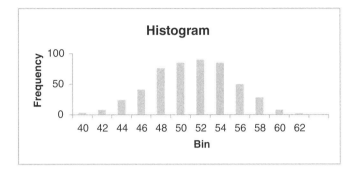

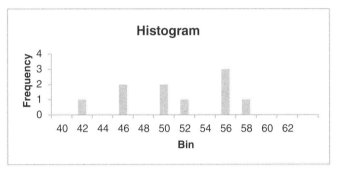

Figure 4.4 Exploratory graph checking for normality with a bin size of 2. The top graph is a plot with 500 points randomly generated and the bottom graph shows the first 10 points from that random sequence. With only 10 points, there is minimal visual support either for or against the data being from a normal distribution. The graph almost looks like a uniform distribution.

One rule of thumb for when to check for normality is to take the sample maximum and minimum and compute those values' z-score, or more properly, t-statistic (number of sample standard deviations that those two values are above or below the sample mean), and compare it to samples' standard deviation using the 68–95–99.7 rule. If the data set has a 3σ event and significantly fewer than 300 samples, or a 4σ event and significantly fewer than 15,000 samples, then you need to be concerned that the data are not normally distributed.

Tests for Normality

The three main tests and some of their benefits and limitations are given below:

Shapiro–Wilk normality test	It requires a minimum of seven values. It is a regression-type test that uses the correlation of the sample values arranged in ascending order against a normal distribution. One drawback

	is it should not be used if several data values are the same (e.g., the value 5 appears several times in the data). It also should not be used for data sets with more than 2000 data points.
D'Agostino-Pearson omnibus test	It requires a minimum of eight values. This test can handle repeated data values. It works by computing the skewness and kurtosis and then analyzing how these differ from the expected values of a normal distribution.
Anderson–Darling test	This measures the fit of the data against an expected distribution and, thus, can be used for testing goodness-of-fit in more than just the normal distribution. The test tends to give more weight to the tails than either the Spapiro–Wilk or the D'Agostino-Pearson tests, so the test is better able to detect nonnormality in the tails of the distribution. One drawback is it should not be used if several data values are the same (e.g., the value 5 appears several times in the data).

Skewed Curves

Skewness is a measure of the asymmetry of a distribution. Few distributions that involve people-based interactions are normal distributed.

- Task time graphs tend to peak with a long tail of longer times. Most people finish with very similar times, but some take much longer. The nature of the task prevents people from finishing to too quickly.
- Academic test scores cluster around A and B grades with a long tail of C through F.

When graphed, values such as task time or academic test scores do not follow a nice bell-shaped normal distribution. Instead, they graph as a skewed distribution with a clearly asymmetrical graph (Figure 4.5).

A normal distribution has the same mean and median value. For a skewed distribution, the mean and median are not equal. A distribution has a positive skew if the tail on the left side is shorter than the right side and the bulk of the values lie to the left of the mean. A distribution has a negative skew if the tail on

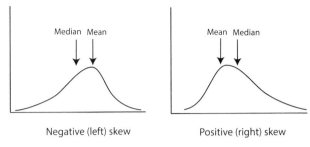

Negative (left) skew   Positive (right) skew

Figure 4.5 Positive and negative skewed distributions.

Figure 4.6 Kurtosis for different distributions.

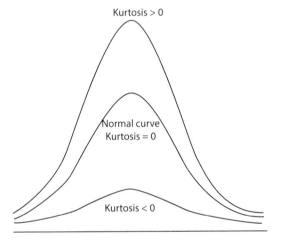

the left side is longer than the right side and the bulk of the values lie to the right of the mean. Figure 4.5 shows an example of both.

Kurtosis is a measure of whether the data are peaked or flat relative to a normal distribution (Figure 4.6). Data sets with high kurtosis tend to have a distinct peak near the mean, decline rather rapidly, and have heavy tails. Data sets with low kurtosis tend to have a flat top near the mean rather than a sharp peak.

Early data analysis must examine the skewness and kurtosis of the data set because most parametric statistical tests assume a normal distribution. Data sets that strongly violate this assumption may need to use nonparametric tests.

Evaluating Skewness and Kurtosis

Statistical software calculates two values that measure the asymmetry of the distribution (Figure 4.7).

The skewness for a normal distribution is zero, and symmetric data should have a skewness value near zero. Negative values for the skewness indicate data

Time		Age	
Mean	182.0133	Mean	28.4800
Standard error	2.7918	Standard error	0.6566
Median	181.0000	Median	29.0000
Mode	192.0000	Mode	31.0000
Standard deviation	24.1781	Standard deviation	5.6863
Sample variance	584.5809	Sample variance	32.3341
Kurtosis	-0.0247	Kurtosis	-0.5681
Skewness	0.2769	Skewness	0.1103
Range	119.0000	Range	26.0000
Minimum	136.0000	Minimum	17.0000
Maximum	255.0000	Maximum	43.0000
Sum	13651.0000	Sum	2136.0000
Count	75.0000	Count	75.0000

Figure 4.7 Skewness and kurtosis value in Excel descriptive statistics. Values for two different variables (time and age) are shown.

that are skewed left (long tail to the left) and positive values for the skewness indicate data that are skewed right (long tail to the right).

By the formal definition, kurtosis for a normal distribution equals three, but many software packages calculate excess kurtosis by subtracting three (resulting in the kurtosis of a normal distribution equaling zero). Mathematically they are equivalent, but the researcher needs to know how the value was calculated to interpret it correctly.

A rule of thumb for evaluating the skewness and kurtosis values is that values between 2 and +2 are considered acceptable to assume the distribution is normal with respect to parametric tests (George & Mallery, 2010). However, depending on the situation and how much they vary as a set (values of skewness = 2 and kurtosis = 2 are very different from skewness = 2 and kurtosis = 0.1). If the values are pushing the limits, it is best to perform also a test for normal distribution.

It is often possible to transform a skewed distribution to get a normal distribution. Section "Transform the Data" discusses data transformation and its limitations.

Effects of a skewed distributions study

Consider a study that looked at two methods of teaching typing. The current method results in students finishing with a mean speed of 42 words per minute (wpm) and with the new method students only typed at 40 wpm. From looking at the mean alone, it would seem that the current method is better or, at best, the two methods are essentially the same. Running a t-test might also reveal no statistical significance.

Figure 4.8 Comparison of skewed curves.

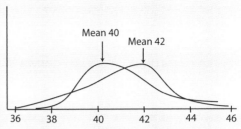

It is important to look at the distribution because skewed distributions can strongly affect the practical significance of the study's conclusions, regardless of the statistical significance of the results. When the results are viewed graphically, the shape of the distribution might tell a different story. The group with the mean of 42 has a heavy tail to the left, which means many people are typing less than 42 words per minute (Figure 4.8). The group with a mean of 40 has a heavy tail to the right, which means many people are typing faster. Something within the study is creating this difference; a researcher needs to work to figure out the cause and what it means for the study goals.

A *t*-test would be the typical statistical test to analyze the two data sets. Because of the highly skewed data, its results may be questionable since it assumes a normal distribution. As drawn, a *t*-test on the data that given by these two lines may or may not give a statistically significant result. But the test only says the data sets are different, it does not say which one is better. Determining the practical significance requires the researcher to make that call based on the studies contextual factors.

Bimodal Distributions

A bimodal distribution has two peaks in its graph (Figure 4.9). A bimodal distribution has equal mean and median as does a normal distribution; without graphing it, the distribution could be mistakenly considered normal. Running statistical tests without taking into account the bimodal distribution will often result in incorrect conclusions.

A bimodal distribution typically comes from the overlap of two different unimodal distributions (only one peak) with different means. For example, in a data set that included both male and female data graphs of factors such as hair length, strength, height, or weight would show two peaks, one for the males and one for the females. Of course, it is clear that these male/female examples would be bimodal, but if the data were collected without gender, it would be more difficult to gain a full understanding of the data. Unfortunately, in most

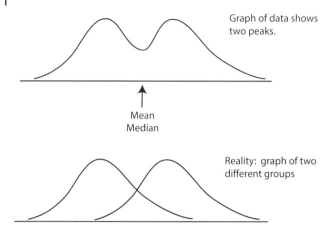

Figure 4.9 Bimodal distribution. The graph shows two peaks. In this example, the overall mean equals the overall median, but only because the two distributions have similar shapes; this rarely happens in practice.

situations, the factors driving the bimodality (in this case gender) are not so evident and the researcher may not initial realize the variable is important. It is a task of the research analysis to realize there may be an overlap of two distributions occurring and to figure out how to separate them out.

Complicating the analysis is that a mixture of two unimodal distributions with differing means does not always appear bimodal (Figure 4.10). The distribution

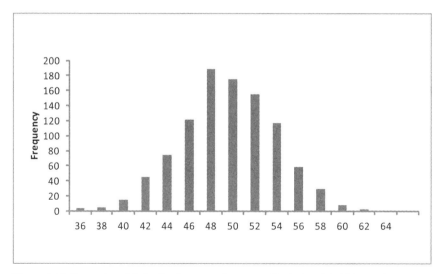

Figure 4.10 Two unimodal distributions too close to easily distinguish. This graph as 500 points in each set with means 47 and 50 with SDs = 4. The graph is somewhat asymmetric, but that it has two populations is not as evident.

of values of two different populations may be expected to show a bimodal distribution, but the difference in mean be too small relative to their standard deviations. The peak may appear as broad flat peak as the two peaks blend or a somewhat asymmetrical graph. The rule of thumb is that a mix of two normal distributions with equal standard deviations is bimodal only if their means differ by at least twice the standard deviation.

When a data analysis fails to find a relationship between variables that are expected to be strongly correlated, the analysis needs to consider whether another study variable (or an uncontrolled variable) is interacting to give a bimodal or multimodal distribution that hides the expected correlation.

Bimodal distribution in coffee drinkers

If a bimodal distribution appears when it is not expected, then there is clearly an unknown factor influencing the study. Follow on research needs to uncover what that the factor is causing the bimodal distribution.

In a study of coffee drinkers (El-Sohemy et al., 2007) found an increased risk of heart attack, but it also showed a bimodal graph like Figure 4.11. Further research found the differences were genetic, with people grouped as slow and fast caffeine metabolizers. Each group had a different set of risk probabilities.

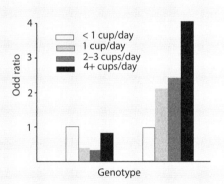

Figure 4.11 Caffeine consumption and heart attack risk. The bimodal distribution is caused by different gene variants people have. (Adapted from El-Sohemy et al. (2007.))

Unfortunately, most of the time, the bimodal graphs are not as pronounced as found in this study. It may require detailed analysis and/or more data—more subjects or different data collection—to uncover the different populations in that what was thought to be a single distribution.

Multimodal Distributions

The bimodal distribution can be generalized to both multiple distributions and multiple dimensions to form a *multimodal distribution* with multiple peaks

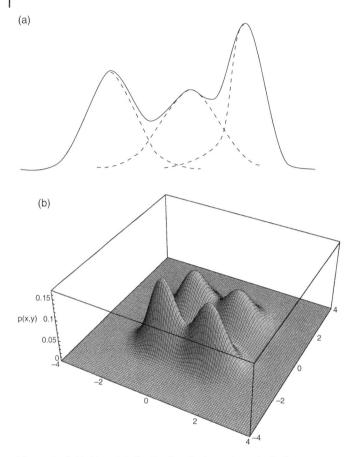

Figure 4.12 Multimodal distribution. Each mode typically forms a separate peak, which the analysis should work to resolve and explain. (a) Shows three overlapping distributions. (b) Shows how the multimodal distribution can extend to three dimensions. (http://en .wikipedia.org/wiki/Bimodal_distribution).

(Figure 4.12). This would be formed when three or more factors act together. As with a bimodal distribution, two modes with insufficient difference in their individual distributions may merge as a single wide peak.

In studies that deal with many interacting variables, the analysis needs to consider whether the collected data are described by a multimodal distribution. Study designs attempt to minimize multimodal distributions by holding variables constant and trying to allow only one to vary, but in social science research, this level of control is difficult to achieve.

Multimodal distributions also helps explain why holding some variables constant and then extrapolating the results may result in very poor predictions. The resulting prediction is based on a combination of factors not relevant to the one being predicted.

Poisson Distributions

The Poisson distribution is a discrete probability distribution that expresses the probability of a given number of events occurring in a fixed interval of time and/or space if these events occur with a known average rate and independently of the time since the last event. That is, it predicts the degree of spread around a known average rate of occurrence.

A Poisson distribution is asymmetrical and skewed to the right, with a zero at one end—there is never a negative event—and a long tail that never reaches zero. As the number of data points increases, the degree of skew diminishes and eventually it will look as symmetrical as a normal distribution. This means that for large data sets parametric statistics can be used because the data is close enough to a normal distribution, but for smaller ones, more likely in social science research, nonparametric statistics may be more appropriate.

Assumptions of the Poisson distribution:

- The event is something that can be counted in whole numbers. A dice roll never gives a value of 3.4; it is always a whole number between 1 and 6. A test answer is right or wrong; there is no partial credit.
- Event occurrences are independent. One occurrence neither diminishes nor increases the chance of another, such as each roll of a dice is independent of previous rolls.
- The average frequency of occurrence for the time period in question is known. The frequency for rolling a 3 on a single dice is 1/6.
- It is possible to count how many events have occurred. The dice were rolled 352 times.

In many study designs, the events are limited to time periods. How many times did the person respond correctly in 30 s? How many firefly flashes occurred in a 10-s time interval? How many questions were answered in 10 min? How many click errors in 5 min?

If a series of events are random and independent, the Poisson distribution can be used to calculate the expected frequencies. On the other hand, in many research studies, the goal is to show that the observed events are not random and independent, in which case showing they violate a Poisson distribution would be part of the analysis.

Suppose someone gets an average of 24 emails per day. That becomes the expectation, but the actual number varies each day: sometimes more, sometimes less. This should fit a Poisson distribution because of the following:

- The email count is a whole numbers. The person cannot receive 22.3 emails in a day.
- Event occurrences are independent. The number of emails received today does not depend on how many were received yesterday or last week.
- The average for the time period is known. In this case, 24 emails per day.
- The total events can be counted. Today, the person received 28 emails and yesterday they received 23 emails.

Given only the average rate, for a certain period of observation, and assuming that the number of emails is essentially random, the Poisson distribution specifies how likely it is that the email count will be 20, or 15, or 31, or any other number, during one period of observation. It could also be used to show that if for 1 day the number of emails was 63, that the chance of that occurring due to random fluctuation is highly improbable. It cannot prove it was impossible, but can claim it was highly improbable and thus due to some nonrandom factor. For example, 34 emails in 1 day have only a 1% chance of occurring. More interesting—and counterintuitive to someone not familiar with probability—receiving exactly 24 emails on any 1 day only has an 8% chance.

Verifying a Poisson distribution in a study

Subjects sit in front of a computer monitor and one of four colored circles randomly flash someplace on the monitor. Depending on the color, they have to perform a different action.

Blue = type "hello"	Yellow = do nothing
Red = click on the circle	Purple = click on the green bar on the screen

The subject's response time to these events should display a normal distribution and that would be the focus of the study's analysis. However, the analysis should also look at the arrival pattern of both time between appearances and color to verify they conform to a Poisson distribution.

Fitting a Poisson distribution shows the prompts were random and independent events, which was an assumption of the study. If they were not random, the subjects may have been able to anticipate the next color, which would decrease reaction time and be a study confound (perhaps they noticed that purple tended to follow blue).

In a laboratory study, the randomness can be relatively easy to achieve. In a natural setting study, although the events may appear to be random, the study analysis should verify that is true. If the study was looking at the number of phone call interruptions at work, the time between calls should be tested to see if they were random.

Binomial Distribution

A binomial distribution describes the probability distribution of binary variables. A binary variable is one which can take one of two possible values—male or female, left- or right-handed, heads or tails, dice roll of 3 or not 3—and the two states of a binary variable are mutually exclusive. It can also be used in studies to describe events such as a task is correct/incorrect or a part was defective/not defective. A graph of a binomial distribution is not 50/50, but would show two different height bars that correspond to the occurrence of the events. Thus a graph of dice rolls for 3 versus not 3 would have one bar five times higher than the other.

The requirement of two possible values can be applies to data sets with multiple values if the data is split into two categories. With dice rolls, it can check that a dice is fair for rolling a 6 with categories of 6 and not 6 (a 1–5 roll). For example, in a study of student grades the two categories could be the following:

- Students who received an A and students who did not receive an A. In other words, everyone else with a B, C, D, and F.
- Students who received an A or B and students who did not receive an A or B. In other words, everyone else with a C, D, and F.
- Testing event by assigning a + or – to each event (it has an attribute or it does not). This can be useful in early data analysis for looking at events as finished task/did not finish task, correct answer/not correct answer.

For large samples, the binomial distribution is well approximated by continuous distributions, and tests such as Pearson's chi-squared test can be used. However, for small samples these approximations break down, and the binomial test should be used. Unfortunately, many studies with small samples use the chi-squared test, since it is probably the test the researcher knows.

Assumptions of the binomial distribution:

- Each trial has only two outcomes. The result is a head or a tail. A dice roll is 3 or not 3 (1, 2, 4, 5, or 6).
- The experiment has n identical trials. The coin was flipped 78 times. A coin can never be flipped 78.3 times.

- Each trial is independent of the other trials. The previous trials do not affect the current trial.
- The probability of getting either outcome is constant. The probability of a dice roll getting a 3 is 1/6 and does not change and the probability of getting a "not 3" is 5/6.
- Includes replacement for each trial. Replacement means the entire population is always sampled. We have 100 data points and randomly sample 10 and calculate the mean. For the next trial, we replace those 10 and randomly sample 10 from the entire set of 100 data points. Thus, some data points may be in both samples.

If the sampling is carried out without replacement, the draws are not independent and so the resulting distribution is a hypergeometric distribution, not a binomial one. However, if the total population is much larger than n, the binomial distribution is a good approximation, and is widely used.

Testing for a binomial distribution can help reveal if the data departs from expectations. For example, the overall population can be considered to have a 50% ratio for male/female. Testing to see if the study sample fits a binomial distribution (perhaps there were 40 females in a random 100 person sample) could let the researcher see if the sample corresponds to the general population.

Cumulative binomial probability refers to the probability that the value of a binomial random variable falls within a specified range. The probability of getting AT MOST 2 heads in 3 coin tosses is an example of a cumulative probability. It is equal to the probability of getting 0 heads (0.125) plus the probability of getting 1 head (0.375) plus the probability of getting 2 heads (0.375). Thus, the cumulative probability of getting AT MOST 2 heads in 3 coin tosses is equal to 0.875.

Binomial distribution calculation example

Binomial distribution tests can show different populations were equally represented in a random sample.

Assume a study for data collected from a random sample of either 100 or 10 people (the numbers for both are listed in the table). Out of the total number, we can look at the number of men or women and determine if the representation fits expectations. We would expect an equal number, but the nature of random samples means there will rarely be exactly 50 : 50 ratio.

Testing for a binomial distribution allows us to check if the actual number still corresponds to the expected 50:50 ratio. If not, then something biased the sample and the study analysis will have to take that into account. Table 4.1 shows the number in the sample and the odds of having that percent or less in the sample. If the percentage is less than 0.05 (like a *p*-value), we reject the

Table 4.1 Binomial distribution percentages.

Total people	100	% ≤	10	% ≤
Women in sample	49	0.460	4	0.377
	42	0.067	3	0.172
	35	0.002	2	0.055

hypothesis that the sample fits the binomial distribution. In this example, only the 35 women out of 100 would be rejected. All of the others can be considered as occurring by random chance.

Notice the difference in the test values between the 100 and 10 total sample column. Even with just 2 women out of 10 in the sample, we cannot reject that it does not follow a binomial distribution, although it is close. In general, testing for a binomial distribution requires a reasonably large number of data points.

Histograms

A histogram is a graphical representation showing a visual impression of the distribution of data (Figure 4.13). The *x*-axis is defined into bins of equal width and plotted with the vertical height equaling the number of events or trials. It provides a researcher with a quick method of accessing the data distribution. Looking at the histogram should be a step in the early data analysis.

Histograms are useful data summaries that convey the following information:

- The general shape of the frequency distribution.
- Symmetry of the distribution and whether it is skewed.
- Modality of the distribution—unimodal, bimodal, or multimodal.

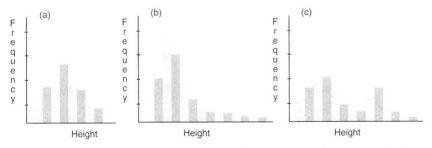

Figure 4.13 Histograms. (a) Approximates a normal distribution. (b) Shows a skewed distribution with a long right tail. (c) Shows a bimodal distribution.

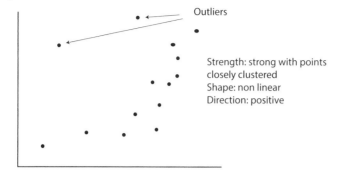

Outliers

Strength: strong with points
closely clustered
Shape: non linear
Direction: positive

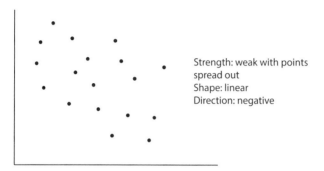

Strength: weak with points
spread out
Shape: linear
Direction: negative

Figure 4.14 Scatter plot with important data analysis points labeled.

Scatter Plots

Scatter plots show the relationship between two variables by displaying interval data points on a two-dimensional graph (Figure 4.14). Its main difference from other graphs is that the data points remain as points and are not connected with lines or areas. They work for interval and ratio data, but should not be used for ordinal or nominal data. In the latter case, the dots will stack up and resemble a bar chart, but without a clear communication component.

Scatter plots are especially useful when there are a large number of data points. They typically plotted with the independent variable on the *x* axis, and the dependent variable on the *y* axis, this provides several important pieces of information about the relationship between two variables:

- Strength
- Shape—linear, curved, etc.
- Direction—positive or negative
- Presence of outliers

Scatter plots work best for two numerical values. A plot that uses ordinal values will end up with stacks of points. In other words, a scatter plot can show

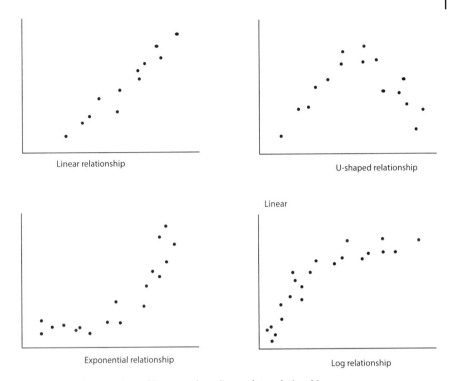

Linear relationship

U-shaped relationship

Linear

Exponential relationship

Log relationship

Figure 4.15 Scatter plots of linear and nonlinear data relationships.

bench press versus weight in pounds, but not bench press versus an ordinal weight value (1 = very underweight, 2 = underweight, . . . 5 = obese).

The graphical display of the relationship between two variables is useful in the early stages of analysis when exploring data. It lets researchers get a feel for the data distribution, so they can determine which type of analysis would be best. It is also good before performing any regression or correlations because it lets the researcher determine if the data requires a linear or nonlinear curve for a best fit (Figure 4.15). The data points require a visual inspection because a linear regression or correlation test will return a value for nonlinear data, but the validity of the results is questionable.

Although scatter plots can show a relationship between two variables, that relationship does not imply a cause–effect relationship. Both variables could be related to some third variable that explains their variation or there could be some other cause. Alternatively, an apparent association simply could be the result of chance or because the data collection occurred over a limited domain.

Box Plots

A box plot (also called a box-and-whisker plot) graphically shows some of the descriptive statistics for a sample.

Box plots are uniform in their use of the box: the bottom and top of the box are always the 25th and 75th percentile (the lower and upper quartiles, respectively), and the band near the middle of the box is always the 50th percentile (the median). The box plot is interpreted as follows:

- The box itself contains the middle 50% of the data. The upper edge (hinge) of the box indicates the 75th percentile of the data set, and the lower hinge indicates the 25th percentile. The range of the middle two quartiles is known as the interquartile range.
- The line in the box indicates the median value of the data.
- If the median line within the box is not equidistant from the hinges, then the data is skewed.
- The ends of the vertical lines or "whiskers" indicate different value. A report that uses box plots needs to define the meaning of the whiskers.
  - The minimum and maximum of all the data (Figure 4.16a).
  - One standard deviation above and below the mean of the data.
  - The 5th percentile and the 95th percentile. Any outliers in the data may then also be plotted (Figure 4.16b).
  - The 2nd percentile and the 98th percentile. Any outliers in the data may then also be plotted.
- The points outside the ends of the whiskers are outliers or suspected outliers.

Box plots are nonparametric and give a visual display without making any assumptions of the underlying statistical distribution. They normally plot the median, rather than the mean; although the mean is sometimes also included. The visual spacing between the different parts of the box helps indicate the range and skewness in the data, and identify outliers.

One drawback of box plots is that they tend to emphasize the tails of a distribution, which are the least certain points in the data set. They also hide many of the details of the distribution. They show the quartile lines, but how the points spread within those boxes are unknown. They might cluster at the median or at the quartile line.

Ranges of Values and Error Bars

Research data always has some level of uncertainty. The data analysis finds a 95% confidence interval for an event as occurring $30 \pm 4\%$ of the time or every 45 ± 4 days. Rather than using $\pm$value, some reports give the values as a range (occurs every 41–49 days). The two different ways are equivalent and does not influence the perceived information quality (Lipkus, Klein, and Rimer, 2001). A research report must be consistent in how it reports the uncertainty levels.

When graphing the data, the error bars (the ± 4 in the previous example) need to be defined: is it the standard deviation, 25%/75% variance levels, the actual observed high/low, or the calculated inaccuracy of the experimental measurements.

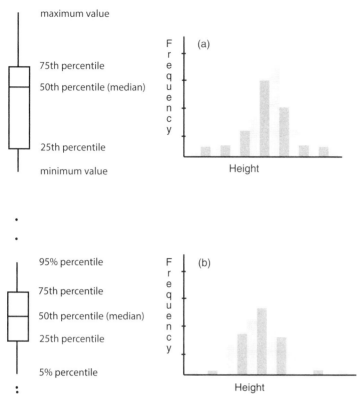

Figure 4.16 Box plots and an associated histogram. (a) Shows data skewed to the left since the median is shifted toward the high numbers. (b) Shows an approximate symmetric distribution (median is centered in the box) and also shows outliers.

The error bars can help readers perceive detailed data patterns (Figures 4.17 and 4.18). In particular, they allow making inferences about the stability of reported data values. Error bars allow readers to mentally impose some segment of a bell curve on to the span of the bar.

Most people lack any experience reading graphs with error bars and they are normally removed as research information is transformed into layman terms. Zacks et al. (2002) found error bars used in about 10% of the journals they sampled, but not in newspapers. Timmermans et al. (2004) question if vertical error bars are a suitable method of graphically presenting information, especially risk information, since it projects a feeling of more complex information. People without experience in evaluating data with error bars tend to ignore the bars and only look at the center point. If the data are already mistrusted, they may see all points within the spread of the error bar as equal and decide the data are suspect. At one level, the mistrust of the data stems from being able to see how precise (or imprecise) the values actually are.

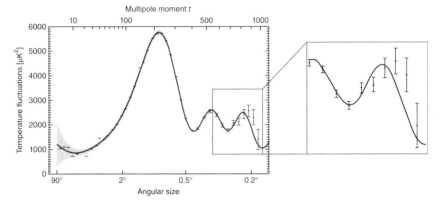

Figure 4.17 Graphics with error bars (the expanded box shows the error bars in greater detail). (NASA/WMAP Science Team.)

Study conditions	Recorded value (mean)	Standard deviation
None	9.11	2.42
Low	9.84	1.91
Medium	9.70	2.13
High	8.85	2.31

Figure 4.18 Error bars on a graph. The data table was converted into a bar graph with error bars showing the standard deviation.

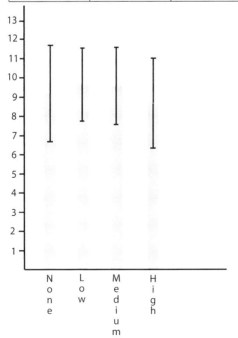

In Figure 4.18, the error bars show the standard deviation of the data around the average. This allows visual method of getting a better feeling for the overall data spread.

References

El-Sohemy, A., Cornelis, M.C., Kabagambe, E.K., and Campos, H. (2007) Coffee, CYP1A2 genotype and risk of myocardial infarction. *Genes & Nutrition*, **2** (1), 155–156.

George, D. and Mallery, M. (2010) *SPSS for Windows Step by Step: A Simple Guide and Reference*, 17.0 update (10a edn). Boston: Pearson.

Lipkus, I.M., Klein, W.M., and Rimer, B.K. (2001) Communicating breast cancer risks to women using different formats. *Cancer Epidemiology Biomarkers and Prevention*, **10**, 895–898.

Timmermans, D., Molewijk, B., Stiggelbout, A., and Kievit, J. (2004) Different formats for communicating surgical risks to patients and the effect on choice of treatment. *Patient Education and Counseling*, **54**, 255–263.

Zacks, J.M., Levy, E., Tversky, B., and Schiano, D. (2002) Graphs in print. In: M. Anderson, B. Meyer, and P. Olivier (Eds.), *Diagrammatic Representation and Reasoning*, London: Springer-Verlag, pp. 187–206.

5

Statistical Tests

This chapter explains the tests used in the data analysis section. The data analysis sections use the test with a sentence or two of justification why it was used, but does not explain how to do the test.

Inter-Rater Reliability

Sporting events such as figure skating have multiple judges, who each gives the participant a score. In high-profile events, there are often claims of biased judging, with one judge consistently scoring lower or higher than the others. Inter-rater reliability tests give a method of evaluating if the judges did agree.

Likewise, many studies collect data that the researcher has to evaluate and assign some sort of qualitative value to, for example

- quality or errors in work, such as grading essays or counting errors,
- quality or type of interaction between people needs to be rated (conversation, aggression, etc.), and
- decisions that require a rating as good/bad.

Note that all three of these list items deal with the researchers and how they evaluate research data; it is not concerned with how subjects performed an evaluation. If a study is looking subject aggression, the inter-rater reliability measures how close two researchers rated a subject's aggression. It is not concerned with the subject's aggression itself.

To improve the consistency of the evaluation, studies typically have two or more people to independently evaluate the items of the research findings. After the researchers perform these rating, the study analysis moves forward using those rating for the analysis. Depending on the study design, it may use each set of ratings, or average them in some manner.

Regardless of how the data are used within the data analysis, the researchers involved with doing the rating must perform them in a consistent manner both

Introduction to Quantitative Data Analysis in the Behavioral and Social Sciences,
First Edition. Michael J. Albers.
© 2017 John Wiley & Sons, Inc. Published 2017 by John Wiley & Sons, Inc.
Companion website: www.wiley.com/go/albers/quantitativedataanalysis

individually and between each other. One evaluator cannot consistently give a score of 2, while the other gives a score of 5 to the same item or items one evaluator scores as a 3 gets values ranging from 1 to 5 from another evaluator. *Inter-rater reliability* provides a measurement for consistency of the rating system. The evaluators are not expected to agree 100% of the time, but the overall consistency of their agreement strongly affects the overall quality of the research results.

Inter-rater reliability is a measure of the ability of two or more individuals to be consistent. Training, practice, and monitoring can enhance inter-rater reliability. Procedures for training the researchers to ensure consistent evaluation must be part of study's initial design. Data analysis needs to start with the analysis of the inter-rater reliability since if the value is low, the evaluation section of the study may need to be modified and redone to obtain acceptable results.

Research results that involve two or more researchers coding data must report the inter-rater reliability. Readers have more confidence in studies with a high inter-rater reliability. Studies with a low inter-rater reliability—the evaluators did not consistently agree—tend to have a low validity.

Inter-rater reliability in a study

A study is looking at the consistency of instructor comments given when grading research papers. Multiple researchers are independently coding all of the research papers in the study. Each instructor comment receives a ranking (from 1 to 5) based on set of criteria the researchers have agreed upon (e.g., does it clearly state what the problem is).

One researcher may give a specific comment a value of 3 and another researcher gives that same comment a value of 4. With values of 3 and 4, the two researchers are consistent and they have a high inter-rater reliability. But what if on another comment, the assigned values were 1 and 5? Here, the inter-rater reliability would be very low.

Of course, looking at a single comparative value of 3 versus 4 has little meaning. But inter-rater reliability looks at all of the comparisons—such as the quality evaluation of all 134 comments recorded in a study. Taken as a whole, it becomes clear if the evaluators were consistent or not.

Calculating Inter-Rater Reliability

The statistics for inter-rater reliability can calculated in several different ways, with different methods appropriate for different types of measurements. The method used for the calculation must be reported in the study write up.

Some ways to calculate the inter-rater reliability are the following:

Joint-probability of agreement	This is a simple calculation of the percentage of agreement between scores. Unfortunately, while simple to calculate, this is also the least robust measure of inter-rater reliability.
	For example, 100 data points were rated by assigning a value of 1–5. The joint probability of agreement can be calculated for each value by calculating the percentage of agreement between the raters. If, for 73 of the 100 data points, both raters assigned them the same value, then the percent of agreement would be 73%.
	Values could also be considered a match if they were consecutive number (or some other predetermined range). Thus scores of 2–3 or 4–5 would be counted as agreeing, but 1–3 or 3–5 would not count as the same value for purposes of calculating percent of agreement. The acceptable variation must be defined before being the data analysis; it is poor research protocol to adjust them during the data analysis to obtain an acceptable agreement value.
Cohen's kappa	Cohen's kappa is probably the most common reported measurement of inter-rater reliability. It treats the data as nominal and assumes the ratings have no natural ordering. Most statistical software packages can calculate Cohen's kappa.
	Cohen's kappa is used for two raters who have rated data points into mutually exclusive categories (i.e., a data point can only receive a single integer value between 1 and 5; it cannot receive 2.4 or a value of both 3 and 4).
	Kappa varies between −1 (perfect disagreement) and 1 (perfect agreement), with 0 meaning pure chance agreement. In general, expected values are between 0 and 1, since a negative value means the agreement was worse than chance.
	A value for an acceptable Cohen's kappa cannot be specifically defined. But to support validity claims for a study, the value should be higher than 0.5. In general, values are ranked: 0–0.20 as poor, 0.21–0.40 as fair, 0.41–0.60 as moderate, 0.61–0.80 as substantial, and 0.81–1 as excellent.

The difficulty with assigning definite value breaks is that it is influenced by outside factors. The value (better agreement among raters) is higher for a small number of codes (assign a value of 1–3) rather than for a large number of codes (assign a value of 1–15). Also, it varies depending on the probability of assigning a code; if most of the items are expected to receive a value of 1 or 2 with 3, 4, and 5 only receiving a few (such as assigning a letter grade to an essay), the value will be different than if values of 1–5 are evenly distributed across the data points.

Cohen's kappa works better than joint probability of agreement because it takes into account the agreement occurring by chance. If values of 1–5 were assigned and if both researchers randomly assigned 1–5, the expected agreement would be 20%.

Fleiss' kappa

Fleiss Kappa is similar to Cohen's kappa, but can be used for 3 or more raters. It varies between 0 (perfect disagreement) and 1(perfect agreement). In general, the value should exceed 0.6 to claim at least moderate agreement. Like Cohen's kappa, it treats the data as nominal and assumes the ratings have no natural ordering. It can be used only with binary or nominal-scale ratings.

Krippendorff's alpha

Krippendorff's alpha measures the inter-rater reliability. Like Fleiss's kappa, it can be used for several raters. It also supports a range of coding values (binary, nominal, ordinal, interval, and ratio). Because of this flexibility, it has a long history of being used in coding of textual analysis, coding open-ended survey questions or discussions, and coding observational studies.

Spearman's correlation coefficient

Spearman's correlation coefficient can be used to measure rank agreement among research coding when the ranking uses an ordinal scale. Spearman's correlation coefficient is a nonparametric test that looks at the overall ranking without making any assumptions about the relationships inherent in the rankings. If more than two raters are observed, the level of agreement for the group is calculated as mean of the values of each possible pair of raters.

Spearman's correlation coefficient looks at the consistency of the rankings of the agreements, not the values themselves. So, using a scale of 1–10, if one evaluator always gave a value three less than the other, the Spearman's correlation coefficient would be high (such as 1–4, 4–7, and 6–9).

Inter-rater reliability

Consider if two judges evaluated several science projects and gave them an overall score of 1–7. Table 5.1 shows the scores and Table 5.2 shows the different inter-rater reliability scores.

The differences in the values arise become they measure different aspects of agreement. Cohen's kappa compares if the evaluators consistently gave the same value. Spearman's correlation coefficient looks at the consistency of the overall ranking—even if they gave the science projects different scores, does the best to worst ranking agree?

Table 5.1 Scores given by two raters on a scale of 1–7.

Rater 1	Rater 2
4	5
6	6
3	6
3	5
7	4
4	5

Table 5.2 Inter-rater reliability scores.

Cohen's kappa	0.1176
Spearman's correlation coefficient	0.402

Neither score shows a high agreement. However, with only 6 values and a scale of 1–7, inter-rater reliability is not highly reliable.

Regression Models

Regression gives an equation that relates the independent variables and their interactions to a dependent variable. It helps you build a mathematical model of what is happening within the system. Regression gives us an equation that can be used to predict the results of future trials. Thus, it only makes sense if the data give values that can be used for a prediction of a dependent variable. A regression does not make sense for something like a 10-question survey about fonts on web pages. There is nothing there that the outcome is trying to predict.

Regressions are useful when you have to use some sort of formula to describe and understand what is happening. If you measure reading times on blocks of 100, 200, and 400 words, it should not be surprising that an ANOVA would reveal statistically significant reading time differences between each group. With the increasing number of words, you would expect that.

A regression lets us determine if there is an equation that describes the reading time. A regression test gives an equation that can be used to model reading times. For example, using the regression equation, we could predict the reading time for 325 words.

A regression equation is really solid for the interval spanned by its values (in this case, 100—400 words). We can probably get realistic values for 75 words or 500 words, but trying to apply it to a 5000-word block of text would be suspect.

Typically, there are multiple values in a regression equation. Reading time is not a simple relationship with total word count. Factors such as font type, font size, and lighting levels also matter. The regression equation (Figure 5.1) could have the three main effect terms (font size, light level, and length) and may also have interaction terms (font size × light level, font size × length, etc.). The regression modeling tests for the significance of the terms and the final equation

Equation from the regression analysis with all terms included. The significance values (*p*-value) are listed below each term.

Reading time = 1.2(font size) + 3.2(light level) + 7.3(length) + 0.30(font size × light level)
                (0.063)   (0.002)   (0.012)   (0.47)

                + 0.59(font size × length) + 1.7(light level × length)
                   (0.14)   (0.085)

Final equation, removing the non-significant terms

Reading time = 3.2(light level) + 7.3(length)

Figure 5.1 Example regression equation for reading time. *p*-Value for each term is in parenthesizes. Only the terms with a significant *p*-value are included in the final equation.

only contain the significant terms. Thus, the final regression equation might only have light level, length, and font size $\times$ length as the terms to predict reading speed. All the others would not be significant.

The final equation has two coefficients (3.2 and 7.3). The relative size of the coefficients cannot be compared—making a claim length is slightly more than twice as important as light level—because the term values differ. Light level and length are measured differently. Light level might range from 50 to 250 and length could range from 500 to 5000.

Parametric Tests

t-Test

t-Tests are used to compare two different groups. The *t*-tests have a few assumptions:

- Used to compare two sets of data.
- The data are interval or ratio data. Do not use a *t*-test on ordinal data.
- The data have a normal distribution.
- Some versions assume the same variance in the two data sets: Welch's *t*-test does not.

Repeated Measures

Repeated measures *t*-test, also known as Student's *t*-test, is used when the data values in the two data sets can be paired up and has equal variance. In this case, paired up means the values are connected somehow, such as pretest and posttest values. It is only applicable to within subjects experimental designs. The repeated measures aspect is that the *n*th value in both the before and after data sets refer to the same person. Both data sets must have the same number of data points.

For example, suppose we measure the test scores on a pretest and posttest after some interaction. It is a repeated measure study because the tests were taken by the same people and the scores pair with each other.

Independent Measures

In between subjects experimental design, clearly none of the values in one data set can be paired up with the values in the other but do have equal variance. For example, the final grades in one section do not pair up with the final grades in another. Likewise, the task completion time for two different groups, or comparing different factors of different groups (i.e., running time of men versus women) cannot be paired. In many situations, the number of data points in each group is not the same.

Welch's t-Test

The previous two *t*-tests both assumed that the two data sets had equal variance. If that is not the case, then Welch's *t*-test should be used instead. It can be used for either repeated measures or independent measures data.

There is a debate in the statistical literature about the use Welch's *t*-test. If the data set is small, then you should use it. On the other hand, for large data sets (not a clearly defined value, unfortunately), the *t*-test is quite robust at handling unequal variance. So, for large data sets, the repeated measures or independent measures *t*-test is acceptable.

ANOVA

An ANOVA test looks at three or more groups of data, unlike the *t*-test that only looks at two.

The ANOVA has a few assumptions:

- Used to compare three or more sets of data.
- The data are interval or ratio data. Do not use an ANOVA on ordinal data.
- The data have a normal distribution.

There are two different variations of the ANOVA, which depends on whether or not the data sets have equal variances. Besides checking the data for normality, it is important to check if all the groups have similar variance.

Follow-Up Tests

An ANOVA test can say if a group is statistically significantly different from the others, but it does not say which one is different. If there are three groups, the ANOVA *p*-value < 0.05 only says that at least one is different, but not which one. For three groups, there are three pairings (1–2, 1–3, and 2–3). Perhaps only 1–3 is statistically significant. An ANOVA would give a significant result, but not tell which pairing was significant.

Follow-up tests (typically *t*-tests or Tukey's HSD test) look at each pairing to determine which differ (Figure 5.2).

To reduce the chance of random fluctuation causing as significant finding, follow-up tests should only be performed if the ANOVA indicates significant. Do not skip the ANOVA and perform follow-up tests on as the first step in the data analysis.

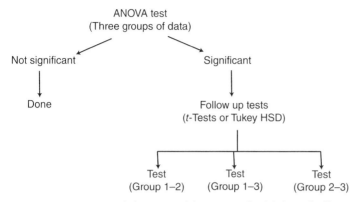

At least one of these tests should show significance
since the ANOVA showed significance.

Figure 5.2 Proper statistical procedure for ANOVA data analysis. Proper methodology of the ANOVA and follow-up tests. Skipping the ANOVA and going directly to the follow-up tests is improper analysis technique.

Nonparametric Tests

When the data sets fail to conform to the requirements for the t-test or an ANOVA, typically with a nonnormal distribution or with noncontinuous data, then nonparametric tests should be used (Blair and Higgins, 1980; Fay and Proschan, 2010; Sawilowsky and Blair, 1992). Unfortunately, there is no firm rule about how nonnormal distributed data should be before a parametric test becomes inappropriate.

Nonparametric tests have less power than parametric tests, but they can handle a wider variation of data distributions. In general, nonparametric tests work with the median values (versus mean values in parametric tests) and work by comparing rankings of the data points. The measurement observations are converted to their ranks in the overall data set: the smallest value gets a rank of 1, the next smallest gets a rank of 2, and so on.

For example, a nonparametric test should be used for two independent samples:

- When the data distributions are asymmetric (that is, the distributions are highly skewed).
- The distributions have large tails (nonnormal distribution).
- The distributions have a uniform distribution (i.e., relatively even number of answers to a Likert-scale question or situations such as testing for a fair dice roll with equal numbers of each value).
- The data set consists of one attribute variable and one ranked variable. Both t-test and ANOVA assume the two variables are the same type.

A nonparametric test exists for essentially all parametric tests.

Parametric test	Nonparametric test
Repeated measure *t*-test	Wilcoxon signed-rank test
Independent measures *t*-test	Mann–Whitney test
ANOVA	Kruskal–Wallis test

If the nonparametric tests return a *p*-value that can be interpreted in the same manner as a parametric test, with a *p*-value is less than 0.05 indicating significance.

One-Tailed or Two-Tailed Tests

Most of the common parametric statistical tests (notably *t*-test and ANOVA) have two different versions: one-tailed and two-tailed. The actual mathematical calculations for each of these tests are different and choosing the wrong one can result in misleading results. To be statically significant, the results must be in the gray area of Figure 5.3, which means the results fall in the top or bottom 5% of the distribution. Sometimes, it makes sense to look at only the top or the bottom.

One-Tailed Tests

A one-tailed test only looks at differences in one direction, either the top or bottom 5% (assuming a *p*-value of 0.05). A one-tailed test implies the researcher knows in advance which mean should be larger; the question is whether it is

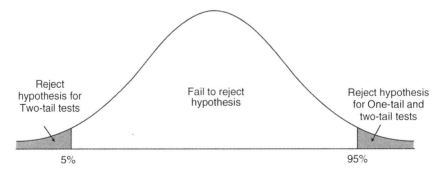

Figure 5.3 One-tailed and two-tailed areas for statistical significance.

significantly larger. The one-tailed tests are only applicable when these two conditions apply:

1) **Changes in the other direction are not of interest to the study design.** For example, we want to know if a new teaching method has significantly improved the student's writing ability over the old method. We consider it equivalent if it made their writing ability remained the same or got worse in a statistically significant way.

2) **There is strong confidence that the change between the two data sets will be in one direction.** For example, we are using a new method that should be better. If we have people lift weights for 12 weeks, we expect them to get stronger since there is no physiological reason to expect them to get weaker. The question is if the change is significant. We consider it equivalent if it made their strength remained the same or got worse in a statistically significant way.

With a one-tailed test, the study question becomes whether the improvement (change in either just the positive or negative direction) was significant. For example, a group runs a timed 1 mile run and then runs 3 miles per day for 3 weeks before doing a second timed 1 mile run. In this case, a one-tailed test would be appropriate to determine if their 1 mile time improved. The researcher is (1) only interested if the times improved and (2) has a strong confidence that times will get better.

Examples of when to use a one-tail test:
- A study examines different sized buttons on a web page. Based on Fitts' Law, we know the smaller buttons will be slower. The question is whether the slowdown is statistically significant.
- A study examines whether seniors do better in a class than juniors in a class within the students' major. There is an expectation of improved performance based on senior status and the research design only wants to examine if they do better; the same or worse performance than the juniors is considered equivalent.
- At study examines if the amount of high school English improves freshman composition writing scores. The group with 2 years of high school English is expected to perform worse than the group with 4 years of high school English; the same or worse performance is considered equivalent.

Two-Tailed Tests

A two-tailed test looks at differences in both direction, both the top and bottom 5%. The two-tailed tests are used when there is no strong reason to expect the groups to differ in a single direction. In general, a two-tailed test requires a larger difference between the two samples to show statistical significance.

In the timed running example, with a two-tailed test, the improvement would have to be larger to show a significant result.

Examples of when to use a two-tail test:
- A study compares a new installation manual design against an old design. The researcher wants to find the new design is better, but has no clear justification for that expectation. Also, it would be important to know if the new design was significantly worse than the old design.
- A study examines whether seniors do better in a class than juniors in a class outside the students' major. There are no expectations of improved performance based on senior status. (Courses within the major could justify using a one-tailed test.)
- A study examines whether males or females perform better in a class. There are no expectations for which group should perform better.
- At study examines freshman composition writing ability based on the amount of high school math. There is no strong justification for expecting students with 2 years of math to write better than those with 4 years of math. Note how this is different from the one-tailed test that looked at writing ability and amount of high school English.

Deciding Between One-Tailed and Two-Tailed Tests

Unless there is a solid line of reasoning that fits both criteria for a one-tailed test, the two-tailed test should be used. Because of the underlying test calculations, a one-tailed test will show significant results for a smaller difference in the mean between the groups. Thus, if a one directional change is not supported by the research design, statistically significant results could be obtained when they are not actually supported.

Early in the study design and before collecting data, the researcher needs to determine what tests will be used to analyze the data. These decisions include whether to use one-tailed or two-tailed tests. *t*-Test results include both the one and two-tailed *p*-values (Figure 5.4). However, it is very poor data analysis technique (and shows a lack of understanding of statistical analysis) to run a test, example the results, and, after seeing only the one-tailed test is significant, to justify using a one-tailed test.

Sometimes the choice of one-tailed or two-tailed tests can both be used during the analysis of the same data. Consider a study that collected the maximum bench press values from two groups of people.

One-tailed tests:

- Comparing the maximum values for males versus females.
- Comparing those who did regular weight training versus those who did not.

(a)

t-Test: Two sample assuming equal variances

	Variable 1	Variable 2
Mean	46.121	46.786
Variance	23.314	10.725
Observations	15.000	15.000
Pooled variance	17.019	
Hypothesized mean difference	0.000	
df	28.000	
t Stat	-0.442	
P(*T*<=*t*) one-tail	0.331	
t Critical one-tail	1.701	
P(*T*<=*t*) two-tail	0.662	
t Critical two-tail	2.048	

(b)

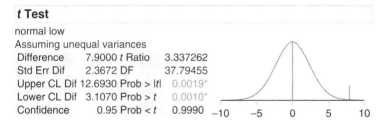

t Test

normal low
Assuming unequal variances

Difference	7.9000	*t* Ratio	3.337262
Std Err Dif	2.3672	DF	37.79455
Upper CL Dif	12.6930	Prob > |t|	0.0019*
Lower CL Dif	3.1070	Prob > *t*	0.0010*
Confidence	0.95	Prob < *t*	0.9990

Figure 5.4 One- and two-tailed *p*-values from (a) Excel and (b) JMP. JMP only shows the *p*-values for the type of test (one- or two-tailed) that was ran.

Two-tailed tests:

• Comparing the maximum values against other factors such as smokers versus nonsmokers or runners versus nonrunner. We have no clear reason for why these groups should be better based on that factor.

Tests Must Make Sense

Test Makes Sense for the Data

The data analysis must make sense for the type of data that is being analyzed. This may seem like a trivial statement, but too often the data analysis reflects the tests a researcher knows how to perform rather than if they make sense for the data.

For example, a study looks at the reading times on blocks of 200, 400, and 600 words. Since it should take longer to read the longer blocks, it should not be surprising that an ANOVA would reveal statistically significant reading time differences between each group.

The problem can occur if a research just asks a statistician to run the tests. There are three sets of data, so the initial impulse to perform an ANOVA. But what is actually needed is a regression. The study should produce an equation showing how the reading times change: is 600 words three times longer than 200 words or is it a different factor?

There are some ways where the analysis could use an ANOVA, but it requires transforming the data. The data analysis could look at "Does the reading time per word unit change?" In this case, the transformation would be to calculate the reading time per 100 words. Then an ANOVA test could be used to see if there is a difference—looking to see if people read 100 words per unit time slower or faster as the text gets longer. In other words, do people read slower/faster when they read longer passages.

Compare Apples to Apples

At study looks a diet plan and weight loss over a 4 week period in both obese females and normal weight females.

A test of the weight loss between the two groups will probably be statistically significant. The obese group will lose more weight per person than the normal weight people. However, the percentage change of body weight may not be significant.

The percentage change of in a factor is usually a better number to use in the analysis than the raw change when the two populations have big differences. Here, it was total body weight. It could also be topic knowledge (a study looks at change in history knowledge between history graduate students and undergrad nonmajors) or strength change (at study looked at people who had worked out consistently for 2 years and people who never did weight lifting).

The important point is make sure the data analysis is using comparable numbers. This may mean calculating a new value (such as the percentage of weight loss). What values are comparable and how they will be calculated are part of the early study design.

Compare the Same Data

A common problem in new researchers is trying to compare different groups of data. All statistical tests must be run on the same type of data.

If a study looked at the number of hours of study for a final and the grade on the final, it would be tempting to do a *t*-test to compare the two numbers

Table 5.3 Hours of study and final grade.

Hours	Grade
4	85
7	92
3	90
5	88

Statistics software will let you do a *t*-test on these two
columns, but the results are meaningless.

(Table 5.3). But one of the groups of numbers are all relatively small numbers
(3–7 hours of study) versus the grade numbers that will probably range from 85
to 92. Even if the sample size was larger, the spread would probably be about
0–10 h of study and 75–100 grades.

The test will show statistical significance, but the result is meaningless. Any
statistical test must use the same type of number for each grouping in the test.
The test cannot compare hours of study against grades. It must be grades to
grades or hours of study to hours of study.

In the data analysis for this study, the hours of study could be used to group
the subjects into two or more groups (such as, low, medium, and high number
of hours of study) and then we can compare the final grade based on those
groups. That would mean the tests are being performed on the same data—the
final grade. Or it could be flipped and the final grades used to create the groups
by hours of study. But either way, the tests must use the same type of data in
each group.

Simpson's Paradox

Simpson's paradox is a statistical effect that shows trends or findings for
different groups that disappear or reverse when the data are looked at differently
(Table 5.4).

A study looked at the income of two different groups.

Table 5.4 Average income by group.

Group 1	Group 2
4896	12,477

Table 5.5 Average income by age group.

Age group	Group 1	Group 2
5–10	300	320
11–15	350	375
16–20	2300	2250
21–25	21,400	21,500
26–30	26,100	25,950

The first round of analysis shows that the two groups have very different income levels.

But if the study looks at the average income by age group, to figure out where the difference actually exists, the differences seem to go away. Table 5.5 shows that each age group has comparable incomes. The 26–30 age group 1 even makes a higher income than group 2. This gives an example of Simpson's paradox. As a whole, the two groups have very different average incomes, but at the closer matched data levels, the difference goes away.

Resolving the paradox requires looking at the number of people in each group (Table 5.6). Then it is clear that group 1 has a much larger number of young people (with very low incomes) than group 2. This shifts the mean and makes the two means look very different.

Data analysis needs to consider a potential of Simpson's paradox confounding the results. In this example, a political group might claim that because group 1 has such a lower income than group 2 government intervention is required. Yet, a closer look at the data does not bear out that conclusion.

In this example, there is the design issue that the two group samples were not comparable. Simpson's paradox can also occur when the totals between the groups are not similar. Ross (2004) looks at the batting average of two baseball

Table 5.6 Number of people in each age group.

Age group	Group 1 people	Group 2 people	Group 1 people	Group 2 people
5–10	48	22	144	90
11–15	54	34		
16–20	42	34		
21–25	22	66	118	134
26–30	10	32		

players and shows how—because of the different number of bats—one player has a higher battering average each year, but the other player has a higher batting average over all the years.

References

Blair, R.C. and Higgins, J.J. (1980) A comparison of the power of Wilcoxon's rank-sum statistic to that of Student's *t* statistic under various nonnormal distributions. *Journal of Educational Statistics*, **5** (4), 309–334.

Fay, M.P. and Proschan, M.A. (2010) Wilcoxon–Mann–Whitney or *t*-test? On assumptions for hypothesis tests and multiple interpretations of decision rules. *Statistics Surveys*, **4**, 1–39.

Ross, K. (2004) *A Mathematician at the Ballpark: Odds and Probabilities for Baseball Fans.* Pi Press.

Sawilowsky S. and Blair R.C. (1992) A more realistic look at the robustness and type II error properties of the *t* test to departures from population normality. *Psychological Bulletin*, **111** (2), 353–360.

Part II

Data Analysis Examples

This section examines the data analysis section of three studies as an independent thing. The data analysis sections use the test with a sentence or two of justification why it was used, but does not explain how to do the test.

Introduction to Quantitative Data Analysis in the Behavioral and Social Sciences,
First Edition. Michael J. Albers.
© 2017 John Wiley & Sons, Inc. Published 2017 by John Wiley & Sons, Inc.
Companion website: www.wiley.com/go/albers/quantitativedataanalysis

6

Overview of Data Analysis Process

The analysis planning should be part of the basic methodology design and never something pushed back until the data have been collected. It is poor practice to collect the data and then begin to determine how you will analyze it.

Data analysis procedures should be planned at the same time as initial study design methodology. The data to collect and what form to collect in it is determined by the desired results. The data analysis needs to be planned to allow reaching the desired results. Thus, the entire process works in reverse: define what the study wants to show, what data analysis methods can give those results, and what type and format of data needs to be analyzed.

Know How to Analyze It Before Starting the Study

Determine the Tests

After developing the study hypothesis and before determining the data collection methodology, you should determine how the collected data will be analyzed to accept or reject the hypothesis. What data to collect and the form in which to collect it is determined by the desired results that come out of the data analysis. In the previous sentence, desired results does not mean whether or not the hypothesis is rejected (one definition of desired result) but by the form of the results. The goal of research focuses on the practical significance and the data collected needs to provide a means of supporting any claims of practical significance. The specific tests and how they will be performed should be determined, not a generic "I will collect data about the groups and then I will do some statistical analysis on the data." In other words, not "I will do a t-test on the task times," but instead, "Group the task times by older and younger than 25 and do a t-test to determine if the age affects task time."

Introduction to Quantitative Data Analysis in the Behavioral and Social Sciences,
First Edition. Michael J. Albers.
© 2017 John Wiley & Sons, Inc. Published 2017 by John Wiley & Sons, Inc.
Companion website: www.wiley.com/go/albers/quantitativedataanalysis

It is important to know how it will be analyzed before collecting the data and before starting the analysis.

- The limitations and requirements of a test can determine the minimum number of subjects. Power analysis should be performed to determine how many subjects to use.
- Knowing how it will be analyzed allows early exploratory analysis. Before all of the data has been collected a round of graphing can help to verify data integrity.
- The nature of the test may require slightly different data to be collected or another type of data to be collected. Parametric tests work best with ratio data; collecting everything on five-point Likert scales limits the study's conclusions.

The data analysis needs to be thought through. Failing to plan often results in running data through every combination of tests in a random search for something (anything) that shows significance. Such a search may eventually find it, but the result may be random fluctuation. Or they may be trivial, such as the times between running 1 mile and 2 miles is statistical significant.

Determine Unit of Analysis

The unit of analysis is the basic unit that the study and test are examining. The unit of analysis varies depending on what you want to study and how you expect to collect the information. It could be people, a class, a family, and a city.

The unit of analysis can be different from the how the data itself are collected. For example, data can be collected at the individual level, such as study habits and writing ability of students across a university. But the unit of analysis might be the following:

- Year in school (freshman, sophomore, junior, or senior) where all of the data for freshman are compared with the other three groups.
- Parent's income level where the data is compared based on income.
- Classroom level unit of analysis if the study compares different teaching methods across multiple sections of the same class, or
- Schedule level that compares final writing ability against the nonwriting-intensive classes taken.

Perform an Exploratory Data Analysis

Fundamentally, the reason for collecting and analyzing data is about detecting and explaining patterns—something which the human eye is very good at.

Although people are good at seeing patterns, they are very poor at seeing patterns in numerical form. Exploratory data analysis helps reveal the patterns that are hidden in numerical data by converting to graphs or other summary groupings. As a result, it gives a basis for interpreting the results of the statistical analysis and determining the practical significance of the findings.

Exploratory data analysis examines the data before crunching numbers looking for statistical significance. Inexperienced researchers want to perform a statistical analysis immediately, but unless the data distributions fit the test requirements, the study can end up with misleading or invalid results. Never, ever, run any statistical test without understanding the data—perform an exploratory data analysis first!

Some of the purposes for doing an exploratory analysis are the following:

- To identify possible errors in the data, e.g. find outliers.
- To reveal features of the dataset. Is it a normal or skewed distribution? How much scatter does the data have? How will the distribution and/or scatter affect the statistical analysis?
- To test for a normal distribution.
- To determine whether parametric or nonparametric tests should be used.

Exploratory data analysis includes the following:

Descriptive statistics (numerical summaries): mean, median, range, variance, standard deviation, etc.

Graphing the data sets, using methods such as
- frequency distribution histograms,
- stem & leaf plots,
- scatterplots,
- box & whisker plots,
- normal probability plots, and
- graphs with error bars.

Perform the Statistical Analysis

Only after completing the exploratory analysis are you ready to perform inferential statistical analysis on the data.

Figure 1.2 in Chapter 1 shows the cyclic nature of data analysis. The planned analysis covers the first cycle of that analysis, but then the researcher needs to take the results of the study and look deeper.

At each point, the statistical results need to be connected back to the study context and the questions of practical significance. These two issues drive the design of the further cycles of the data analysis.

Analyze the Results and Draw Conclusions

After completing the inferential statistical analysis, those results must then be interpreted. Knowing the ANOVA gave $p = 0.023$, thus the results are significant is not adequate. The purpose of research is to explain why the groups are different and to understand the underlying factors that made them different; both of which contribute to the practical significant of the results. Thus, the entire study situational context and overall data set distributions need to be considered as part of explaining what that value of $p = 0.023$ means for the study's research questions. Reporting the p-value and stopping is insufficient and poor research.

Interpreting the analysis results can be visualized as a funnel, with the specific tests and interpretations of the data variables at the top and narrowing down to a combined summary focused on the research question (Figure 6.1). It is important to remember that any summary results in loss of information. The fullness of the original data has been cut down to the essentials. However, the summary does provide a razor focus on the essential data, the points the researcher has deemed most important about this data set. As a result, the important aspects have a high salience since the extraneous data has been removed.

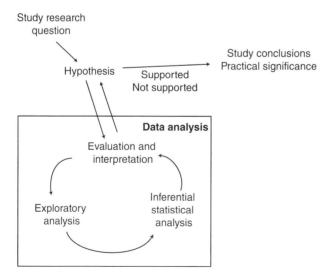

Figure 6.1 Result interpretation cycle. The results of statistical tests lead to conclusions about the practical significance.

Writing Up the Study

After completing the analysis and interpreting the results, then you can write up the final study report.

Reporting Statistics in Research Reports

When reporting quantitative results, both the statistical significance and effect size should be reported. Since 1994, the American Psychological Association has been officially encouraging authors to report effect sizes when reporting data.

Thus, study reports should include statistical significance, effect size, and confidence intervals in the results section and the discussion section should consider the relationship of these three values to the big picture in which the study is situated. For example:

Found significance	Student's spend different amount of times studying for the exam. There was a statistically significant difference between two groups with $p < 0.02$. Therefore, we reject the null hypothesis that there is no difference in study time for exam scores. Further, Cohen's effect size value ($d = 0.62$) suggested a moderate to high practical significance.
Found no significance	Student's spend different amount of times studying for the exam. There was a no statistically significant difference between two groups with $p = 0.24$. Therefore, we fail to reject the null hypothesis that there is no difference in study time for exam scores. Further, Cohen's effect size value ($d = 0.09$) suggested a low practical significance.

Unfortunately, in colloquial English the word "effect" implies causality but the statistical term "effect size" does not. However, many readers may still make the subconscious association of effect = cause. When writing up reports, researchers have consider their audience to avoid having their readers draw unintended conclusions from the text.

Verbal and Numerical Probabilities

The write up of a data analysis often uses both the numerical probabilities derived as part of the analysis (73% of the subjects performed at excellent or higher) and verbal descriptions (a high chance of the subjects performing

Table 6.1 Verbal probability expressions and their numerical equivalents (adapted from Mazur and Hickam, 1991, p. 239).

Word	Mean probability (%)	Word	Mean probability (%)
Almost certain	86	Possible	49
Very likely	77	Not unreasonable	47
Likely	64	Improbable	31
Frequent	64	Unlikely	31
Probable	60	Almost never	15

better). Likewise, study instruments, such as Likert scale questions, use a verbal scale (strongly disagree to strongly agree). How people mentally translate between verbal probabilities and numerical probabilities needs to be considered in both data analysis and the study report.

The numeracy levels of the readers matter. Gurmankin, Baron, and Armstrong (2004) found that less numerate individuals wanted verbal risk information rather than numeric risk information: They preferred hearing a low chance of rain rather than a 20% chance of rain. People with high numeracy are more likely to retrieve and use appropriate numerical principles and also tend to avoid framing effects. People with lower numeracy are more influenced by irrelevant information, probably because they draw less meaning from the relevant numbers (Peters et al., 2006).

If told there is a low chance of X, how do people mentally translate "low" into a probability? Mazur and Hickam (1991) looked at how patients translate the verbal expressions for the chance of side effects for a medical procedure (Table 6.1). Although they found a spread of actual numerical values for each term, the verbal term order was very consistent. So, although people may have different mental values for "likely," they consistently put "likely" as higher than "frequent."

References

Gurmankin, A.D., Baron, J., and Armstrong, K. (2004) The effect of numerical statements of risk on trust and comfort with hypothetical physician risk communication. *Medical Decision Making*, **24**, 265–271.

Mazur, D. and Hickam, D. (1991) Patients' interpretations of probability terms. *Journal of General Internal Medicine*, **6**, 237–240.

Peters, E., Västfjäll, D., Slovic, P., Mertz, C., Mazzocco, K., and Dickert, S. (2006) Numeracy and decision making. *Psychological Science*, **17** (5), 407–413.

7

Analysis of a Study on Reading and Lighting Levels

Lighting and Reading Comprehension

Study Overview

A study used a within subject design [defined on page 199] to examine reading comprehension with different lighting conditions. The subjects read a text and then answered a set of questions about it. Because it was within subjects, two different, but similar, texts were used and they were cross-balanced with the subjects.

The number of correct answers was analyzed to determine if light levels affect comprehension and reading time. The subjects were also videoed and two researchers scored the subjects 1(low) to 5(high) as a ranking of the amount of difficulty they exhibited reading the texts. They looked at factors such as holding the paper close, squinting, and constantly changing paper distance.

Know How the Data Will Be Analyzed Before Starting the Study

Reading time (in seconds) is interval data and the values for test scores and reading difficulty are ordinal data. The lighting conditions are nominal data [defined on page 202]. The test scores would be ratio data, but since they only range from 1 to 5, the argument can be made to consider them as ordinal. In general, that choice should not affect the data analysis.

We will be looking at the distribution during the exploratory data analysis to make a final determination between parametric and nonparametric tests.

Introduction to Quantitative Data Analysis in the Behavioral and Social Sciences,
First Edition. Michael J. Albers.
© 2017 John Wiley & Sons, Inc. Published 2017 by John Wiley & Sons, Inc.
Companion website: www.wiley.com/go/albers/quantitativedataanalysis

Units of Analysis

Independent and Dependent Variables

The independent variables [defined on page 197] are the ones that you want to manipulate. The lighting conditions are the independent variable, since it is the factor the researcher was changing.

The dependent variables[defined on page 198] are the variables that change as a result of the experimental manipulation. The reading time and the comprehension test scores are the dependent variables, since they are both hypothesized to change based on changes in the lighting conditions.

Primary Relationships to Explore

Lighting condition versus reading time	Lower lighting should result in increased reading times.
Lighting condition versus comprehension	Does the lighting level affect comprehension. This would imply that having to exert more mental effort to read the text results in a different level of learning.

A relationship that must be carefully considered before doing any analysis is low lighting correct answers versus normal lighting correct answers. With the small range of values and the questions coming from different texts, it can be difficult to justify they represent equivalent things. Inferential statistical tests are testing if the samples come from the same population; if the study design sets the samples up as different populations, then doing statistics on them is misusing the tests [see page 99].

Secondary Relationships to Explore

Reading time versus comprehension	This would be a factors where the lighting level and reading time may show a two-way interaction [defined on page 50].
Reading time versus difficulty	The reading time should be related to the difficulty the subjects had reading the material.
Comprehension versus difficulty	The comprehension should be related to the difficulty the subjects had reading the material.

This study did not collect demographics or reading ability data on the subjects. A formal study should have collected this data and it would need to be considered within the analysis. In particular, the subject's reading ability will clearly be related to both reading time and comprehension. Depending on the spread of demographic data, such as age or education, those factors could also be considered as part of the analysis.

Ensure There Are No Order Effects

Each subject reads two texts with each text at different light level. There is a chance that the order of reading the texts affected the results [defined on page 200]. In other words, reading text 1 (low light text) first may influence how well a subject comprehends text 2, or vice versa. To control for this, the lighting conditions are reversed for half of the subjects: half start with low light and half start with normal light.

The analysis needs to ensure the reading order did not influence the answers.

We need to check if there are statistical differences between the group that read text 1 first and the group that read text 2 first. The goal is that there will be no difference.

If there is no difference then they can be grouped for the study analysis.

If they are statistically different, then there are reading order confounds that will have allowed for in interpreting the analysis results.

Perform an Exploratory Data Analysis

With an exploratory data analysis [explained on page 108], you need to take the time to examine the data and develop a feel for how the numbers relate to each other. This gives you a basis on judging both the results of the inferential statistical tests and how those results connect back to the study's hypotheses. It also may result in seeing potential or interesting connections that deserve further analysis but were not part of the original data analysis plan. If you jump directly into inferential statistics, you risk finding significance but not having a deep enough understanding of the data to determine what that significance means [see page 34].

Descriptive Statistics

An exploratory data analysis typically starts with descriptive statistics (Table 7.1).

A visual inspection can reveal if any of the numbers seem to be different from the others. The following can be seen in this table.

Correct answers   Correct answers for the low lighting condition mean = 3.95 and SD = 1.19. Correct answers for the normal lighting condition mean = 4.50 and SD = 0.76 [mean is defined on page 37 and SD is defined on page 39]. The correct answers for normal lighting are higher, indicating the comprehension may be better. Also, the SD for normal lighting is smaller, which means the answer spread is tighter around the mean than the low lighting conditions answers.

Table 7.1 Descriptive statistics.

	Low lighting				Normal lighting			
	Correct answers	Time	Rater 1	Rater 2	Correct answer	Time	Rater 1	Rater 2
Mean	3.95	233.30	2.50	1.80	4.50	241.20	3.30	3.75
Standard error	0.27	1.61	0.25	0.26	0.17	1.73	0.23	0.31
Median	4.00	233.50	2.00	1.00	5.00	242.50	3.00	4.00
Standard deviation	1.19	7.20	1.10	1.15	0.76	7.76	1.03	1.37
Sample variance	1.42	51.91	1.21	1.33	0.58	60.17	1.06	1.88
Kurtosis	2.26	0.66	−1.26	2.08	−0.04	0.02	−0.02	−0.40
Skewness	−1.56	−0.58	0.13	1.58	−1.19	0.28	−0.36	−0.86

The correct answer values can range from 0 to 5. Although it is ratio data, the small range makes most descriptive statistics not to than useful. Part of the problem is the difficulty of claiming equal scores are really equal; that assumes all questions are equally difficult and have equal chances of being correct.

The SD values are not similar; the normal lighting condition is only about 2/3 the value of the low lighting condition. This may affect the choice of which t-test to run. t-Tests (and ANOVAs) have different tests depending on if the variance between the samples is equal or different [t-tests are explained on page 93 and ANOVA is explained on page 94]. Remember that SD equals the square root of the variance. The definition of when the variance [defined on page 39] is unequal in many situations is very soft. There are formal tests, such as Levene's test, Bartlett's test, or the Brown–Forsythe test; however, they are not too reliable on small data sets.

With the small data range of 1–5, the skewness and kurtosis have minimal meaning.

Time
Reading time for the low lighting condition time = 233.30 and SD = 7.20. Reading time for the normal lighting condition time = 241.20 and SD = 7.76. As with the correct answers, the normal lighting time reading was faster,

which would be expected. We can also see that the SD values are very similar, so we can use tests with equal variance.

The skewness [defined on page 66] and kurtosis values show that the curve should show a relatively normal distribution. Skewness and kurtosis [defined on page 69] values of less than 2.0 can be considered close enough to normal. A normal distribution would be expected since reading times should be normally clustered around the mean.

Rater

The rating values are ordinal data [defined on page 202], which limits the value of the descriptive statistics. Since this was scored with a Likert scale of 1–5, the use of both median and mean values has to be used carefully. Rater 1 gave similar scores in both conditions and rater 2 gave higher values for the normal light condition. Comparing the two raters, we see rater 1 gave higher scores in the low light condition and rater 2 gave higher scores in the normal light condition. To really understand the rater data, we will have to graph it.

We have to be concerned with inter-rater reliability measures [defined on page 87], since both raters should give the same value to each subject.

Assigning values to nominal and textual ordinal values

For calculating the difficulty, the rater assigned values starting at 1.

Mathematically and statistically, there is no difference between data analysis conducted on five-point scales that use 0–4, 1–5, or −2 to +2 for the assignment of numbers to scale points. They will have exactly the same mean differences between populations, the same standard deviations, and consequently, exactly the same outcomes for any statistical test (parametric or nonparametric) and the same confidence intervals.

Of course, the actual means will be different, but then the mean as a single value has no real meaning. It is only when comparing means that they take on meaning.

Graphical Exploratory Analysis

Graphing the variables gives us a visual indication of the overall spread of values for each of the variables.

Do not overinterpret the analysis results

It can be easy, especially for new researchers, to overinterpret the results or to start to justify/rationalize values. Follow the data analysis process and interpret the results with respect to the situation. Do not start out with a set of assumptions.

For example, if a study has age as a variable, do not start justifying why the older people are slower. They may be, but that is not a claim that comes out of the descriptive statistics.

First you need to follow the data analysis process. If you already know older people are slower, why do the study? It is not that their mean time was 5 s slower; you need to determine if that different is both statically and practically significant. Or, if you need to know "how much slower," then any rationalization is meaningless since it saying what is already known.

Reading Time

Reading time should show a something close to a normal curve [defined on page 64], since some people's reading time should be around the average, with some faster, and some slower. However, there is a minimum time, so the curve will not really be normal. However, it should be close enough for parametric tests.

Creating these histograms in Excel

The exploratory analysis in this example, and in the other two long examples, contains several multibar histograms [defined on page 79], such as Figure 7.1. Unfortunately, Excel cannot create these as a single step.

1) Using the data analysis add-in, create a histogram for each data element. In Figure 7.1, you would create one for normal condition values and one for low condition values.
2) Take the data table output from the histograms, add in the bin values, and use those values to create a bar chart. The bin values are used for the *x*-axis.

Figure 7.1 shows the reading time curve does show a normal distribution. Later, we will do a formal statistical test for normality, but with the low number of data points, it will difficult for a formal test to reject it. So, we proceed with the assumption the reading time data fits the normality criteria for parametric tests.

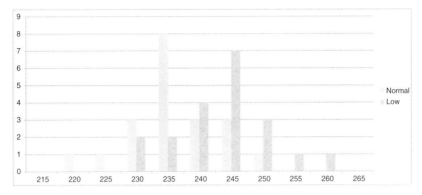

Figure 7.1 Reading time distributions.

The normal light reading times are faster than the low light reading times. We also can see that the reading times for the two conditions are plotting into what looks like a bimodal distribution. This is what we would expect in they have statistically significant differences.

To check for reading order effects, we need to graph the reading time for low light read first versus read second and the normal light read first and read second (Figure 7.2). We want to find highly similar graphs.

The subjects that read the normal light text first seemed to read faster, however almost everyone was within 5 s of the mean. The large apparent differences may be an artifact of the graph scale. During the inferential analysis, we will do a *t*-test to determine if the two sets of time are really different.

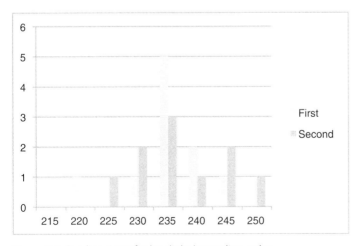

Figure 7.2 Reading times for low light by reading order.

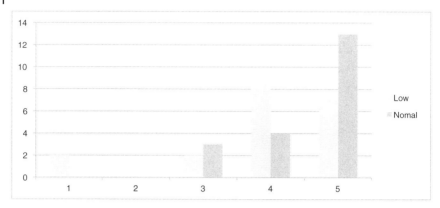

Figure 7.3 Correct answers by light level.

Correct Answers

The study uses the number of correct answers on a posttest to determine the reading comprehension level.

Figure 7.3 shows, as expected, the low light levels appear to cause a lower comprehension. During the inferential analysis, we will verify whether this apparent different is significant. Both curves are strongly not normal curves [defined on page 64], so parametric tests should probably not be used [explained on page 93].

The graph makes it obvious that most of the subjects scored a 4 or 5 on both comprehension tests. Unfortunately, this is indicative of a ceiling effect [defined on page 182]. This will make it more difficult to obtain statistical significance and weakens any study claim.

Ratings Values

To check the ratings, we have a couple of different choices. The two rater's values can be summed or averaged. This analysis uses the sum (Figure 7.4). Since the exploratory analysis's purpose is to get a feel for the data, the choice is up to the researcher.

The normal lighting condition clearly has much higher overall rater values than the low lighting condition. We would expect people to be able read better with normal lighting, so this fits expectations.

We also need to check if the two raters were consistent in their evaluation of the reading difficulty (Figure 7.5) [explained on page 87]. We will check the agreement with a graph that looks at the difference of the two rating values: Rater 1 value is subtracted from rater 2 and that value graphed.

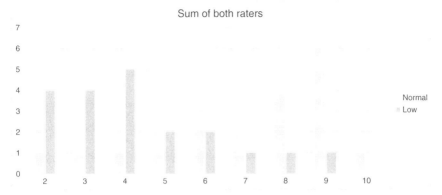

Figure 7.4 Ratings by light level. The values plotted were the sum of the two rater's values for each subject.

Ideally, the value will always be zero. The negative values occur when rater 2 gives a higher value.

The zero bar is the highest in the low light condition. It is similar to the size of the −1 bar in the normal condition. Since almost all of the bars are the −1 to 1 range, we can see that raters did almost always agree within one number.

The graph structure reflects the descriptive statistics that found rater 1 gave higher scores in the low light condition (resulting in positive values) and rater 2 gave higher scores in the normal light condition (resulting in negative values).

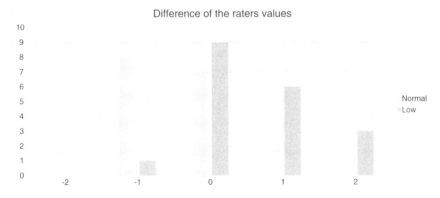

Figure 7.5 Rating differences for the low light level condition. The value plotted for each subject is equal to rater 1 − rater 2.

We could also graph the pair of rater values for each subject to visually examine how consistent the ratings are. With 20 subjects the graph is barely manageable; it will not work for larger subject pools.

Summary of Exploratory Data Analysis

Having completed the exploratory data analysis, we can draw these conclusions and have these expectations as we proceed into the statistical data analysis.

• The reading time for the normal condition had a lower mean, as expected. We need to test if the difference is significant.
• There does not appear to be an order effect for the reading time based on reading order. This is a desired result and means we do not have to worry about order when comparing low to normal light conditions.
• Correct answers, which we are using as a measure for comprehension, has a ceiling effect. Normal light condition was higher, but the ceiling effect may mask any significance.
• The two raters both found that people had less difficulty in the normal light condition. However, there was a wide range of difficulty values, so this may reflect subject reading ability (which was not controlled for) instead of the lighting condition.

Perform an Inferential Statistical Analysis

Having completed the exploratory data analysis, we are ready to proceed into doing an inferential statistical data analysis [explained on page 6].

Testing for Order Effects in the Reading Times

First, we need to determine if there is an order effect [defined on page 183] caused by reading the low light or normal light text first. From the exploratory analysis, we do not expect to find a difference, which is the desired result.

There can be an argument made both ways for whether this should be a t-test (parametric) [defined on page 93] or a Mann–Whitney test (nonparametric) [defined on page 95]. Both tests will be run here and the results compared. Mann–Whitney test was ran online at http://www.socscistatistics.com/tests/mannwhitney/Default2.aspx

In both cases, the p-value [defined on page 19] from both the t-test and the Mann–Whitney test is nonsignificant, which is our desired result (Table 7.2). This means the reading order did not create a confound in the study.

Table 7.2 Comparisons of for reading order.

t-Test: Two-sample assuming equal variances			Mann–Whitney results
	First	Second	The *U*-value is 46.5. The critical value of *U* at $p \leq 0.05$ is 23. Therefore, the result is not significant at $p \leq 0.05$.
Mean	232.700	233.900	
Variance	47.789	60.989	
Observations	10.000	10.000	
Pooled variance	54.389		
Hypothesized mean difference	0.000		
df	18.000		
t Stat	−0.364		
P(*T* < = *t*) one-tail	0.360		
t Critical one-tail	1.734		
P(*T* < = *t*) two-tail	0.720		
t Critical two-tail	2.101		

Both tests give a nonsignificant result (two-tailed *p*-value > 0.05), which is the desired result.

What if there was an order effect?

In this example, the tests for differences based on reading order are not statistically significant. We wanted them to be not significant, so we would not have to allow for that in the data analysis. However, in some similar studies, this will not be the case. So, let us consider how the analysis would be different if the results had been significant.

As a first step, rather than running a single comparison tests (such as word proficiency versus difficulty), run three: one for combined values and one for each order condition.

- If all three are significant or not significant, then the issue of differences between sections does not matter.
- If one order condition is significant and the other is not, then you have to delve deeper to attempt to uncover how and why the differences between the reading orders are causing the differences. This may require going back and gathering additional data to increase the sample size.
- There is also the 5% probability (a *p*-value of 0.05) that the results occurred by chance. Actually, at this point, the overlap of the 95% confidence intervals becomes important. Clearly, there is a difference in the judgment call when the *p*-value is 0.045 or <0.001.

Further thoughts on the comparison of the graphical data and statistical analysis

Sometimes the exploratory analysis gives a graph where it seems the two samples should be different. For example, Figure 7.6 shows a graph of word proficiency for the students in two different class sections. Visually, they appear different, but statistically, they are the same. This helps to emphasize that a visual examination of the data does not substitute for a rigorous evaluation. The small sample size also makes it more difficult to determine if they really are different.

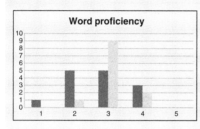

t-Test: Two-Sample Assuming Equal Variances

	Section 203	Section 206
Mean	2.714	3.083
Variance	0.835	0.265
Observations	14.000	12.000
Pooled Variance	0.574	
Hypothesized Mean Difference	0.000	
df	24.000	
t Stat	−1.238	
P(T<=t) one-tail	0.114	
t Critical one-tail	1.711	
P(T<=t) two-tail	0.228	
t Critical two-tail	2.064	

Figure 7.6 Word proficiency between two sections.

Section 206 has a much higher bar at 3 while the section 203 has a wider spread of values. There are two different things happening here that any researcher must consider.

The limited data set size makes it difficult to get statistical significance. Remember that statistical significance is conceptually based on having one set of values fall outside of the 95% confidence interval of the combined data set. Small data sets have very wide confidence intervals.

The limited number of values and tight cluster of values (2, 3, or 4) give very little spread of the data. *t*-Tests assume interval data, not ordinal. The small range of possible values means the standard deviation is small. The Mann–Whitney test is ranking the values, but it also has difficulty when the values have little spread. There is not much to rank.

On the other hand, this difference in graphical representation must be kept in mind when interpreting the results. Differences that are found visually may help to explain some of the differences that are found statistically. It is never enough to simply state statistical significance or nonsignificance. The goal of a research study is to provide a conceptual explanation for that statistical finding and, as such, the study needs to justify that the statistical finding makes sense with respect to the data. This is also why a researcher cannot simply hand the data to a statistician, receive back a set of results of significance or not significance, and write up those results.

Test for Reading Time Versus Lighting Condition

The data from the two conditions can now be analyzed to determine if the independent variables are related to the dependent variables.

A primary relationship study explored was whether the lighting influenced the reading time. We can expect the reading time to be slower in the low lighting condition and as seen in Figure 7.7, that result is supported. The p-value $= 0.0019$, which is highly significant.

(a)

t-Test: Two-Sample Assuming Equal Variances

	Time low	Time high
Mean	233.300	241.200
Variance	51.905	60.168
Observations	20.000	20.000
Pooled Variance	56.037	
Hypothesized Mean Difference	0.000	
df	38.000	
t Stat	−3.337	
P(T<=t) one-tail	0.001	
t Critical one-tail	1.686	
P(T<=t) two-tail	0.002	
t Critical two-tail	2.024	

(b)

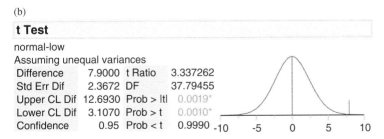

t Test

normal-low
Assuming unequal variances

Difference	7.9000	t Ratio	3.337262
Std Err Dif	2.3672	DF	37.79455
Upper CL Dif	12.6930	Prob > ltl	0.0019*
Lower CL Dif	3.1070	Prob > t	0.0010*
Confidence	0.95	Prob < t	0.9990

Figure 7.7 *t*-Test for reading time versus lighting condition. Result (a) is from Excel and Result (b) is from JMP. Both show the same *p*-value, which is expected.

Test for Comprehension Versus Lighting Condition

A main factor in the study was to see in the comprehension of a text changed when the lighting changed. We can run a test comparing the two sets of answers (low light answers versus the normal light answers).

It is also a highly skewed graph [defined on page 68], so a *t*-test may not appropriate.

To see the differences, we will perform both a *t*-test and a Wilcoxon test in JMP (Figure 7.8). We can compare any differences. Both tests return a *p*-value that shows the comprehension is not affected.

Interpreting this result requires considering that the exploratory analysis (Figure 7.9) shows that there is a potential ceiling effect [defined on page 182] with most subjects getting 4 or 5 correct answers in both conditions. With the clustering of the answers, it is difficult for a test with only 10 subjects to show significance. It has 10 subjects, rather than 20, since each group—read text first and read text second—only has 10 people.

Another way to analyze the comprehension values is to consider them as paired values. This looks at the pair of answers given by each subject (the low light answer and the normal light answer) to determine if they are different. Pairing the data

(a)

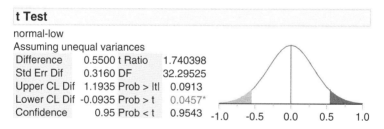

(b)

Wilcoxon / Kruskal-Wallis Tests (Rank Sums)

Level	Count	Score Sum	Expected Score	Score Mean	(Mean-Mean0)/Std0
low	20	352.500	410.000	17.6250	-1.683
normal	20	467.500	410.000	23.3750	1.683

2-Sample Test, Normal Approximation

S	Z	Prob>|Z|
467.5	1.68319	0.0923

1-way Test, ChiSquare Approximation

ChiSquare	DF	Prob>ChiSq
2.8830	1	0.0895

Figure 7.8 (a) *t*-Test for comprehension versus lighting condition and (b) Wilcoxon test for comprehension versus lighting condition.

Figure 7.9 Paired *t*-test. The one-tail test is showing significant and the two-tailed test is showing not significant.

	Low	Normal
Mean	3.950	4.500
Variance	1.418	0.579
Observations	20.000	20.000
Pearson Correlation	0.087	
Hypothesized Mean Difference	0.000	
df	19.000	
t Stat	-1.814	
P(T<=t) one-tail	0.043	
t Critical one-tail	1.729	
P(T<=t) two-tail	0.086	
t Critical two-tail	2.093	

like this may be questionable. We would be comparing the value for two different tests; the low light and normal light texts were different. This pairing only makes sense if the researcher can justify that the answers are comparable.

A researcher needs to examine the data consider how to look at it multiple ways—not just running a single test. This examination should be part of the data analysis plan developed during the study design phase and not pushed back until data analysis.

One-tailed test and post hoc justification

Methods for robust data analysis discourage justifying one-tailed test post hoc. In other words, when the results are nonsignificant for a two-tailed test and significant for a one-tailed test [defined on page 96], they do not justify why to use the one-tailed test after you have test results.

The one-tailed test is significant with $p=0.043$ and the two-tailed test $p=0.086$. Remember the one-tailed test only looks at the upper 5% of the curve and the two-tailed test looks at both ends. A one-tailed test would be reasonable if (1) we only cared to show that lighting only positively/negatively affected the results and (2) there were strong experimental reasons for us to predict a positive/negative increase in the relationship.

Statistical tests require comparing the same data

It seems obvious that the two samples for the test should be the same data (reading time, time to complete and action, before/after test scores). The columns of numbers brought into the test must be the same measurement; otherwise running a statistical test on them is invalid. [For an extended discussion, see page 99.]

Example 1

Some researchers run statistics tests on different sets of data. In this study example, they would do a *t*-test of low lighting correct answers and low lighting reading time. In other words, comparing a series of 1–5 numbers against a series with an average of 233. Any time two sets of numbers like that are compared, they will be significantly different.

Example 2

Some researchers fall into a trap of comparing data with the same range, but which is not comparable.

For another example, the answers on pretest during a health literacy study were scored −1, 0, +1 (incorrect, partially correct, and correct). The posttest answers were scored −1, 0, +1 (worse answer, not improved, improved). The scoring method is part of the data collection methodology and can be viable.

But, these are two different sets of data and cannot be statistically compared. The pretest is correctness of the answer and the posttest is the improvement. They measure different factors. If a subject had an incorrect answer (−1) on the pretest and still had it incorrect on the posttest (0), the data values are different, but the subject still does not know the answer.

In addition, with only three ordinal values such as this example used, a *t*-test is not appropriate. Nonparametric tests should be used.

Test for Reading Time Versus Comprehension

The study did not control for reading ability, so the lack of a relationship between answers and reading time is not surprising. The ceiling effect on the answers also makes a relationship difficult to find.

Both graphs show a very flat line that indicates a low relationship. Coupled with the ceiling effect of the correct values clustered at 4 or 5, the study can make minimal claims about how reading time affects comprehension (Figure 7.10).

If the study had collected more fine-grained data, such as reading ability and used a test that did not show ceiling effects, it would be possible that to show two-way (or more) interactions within the data. [Defined on page 50].

Figure 7.10 Reading time versus number of correct answers. The graph (a) is the low light condition and the graph (b) is the normal light condition. The mostly flat line in both graphs indicates a low relationship.

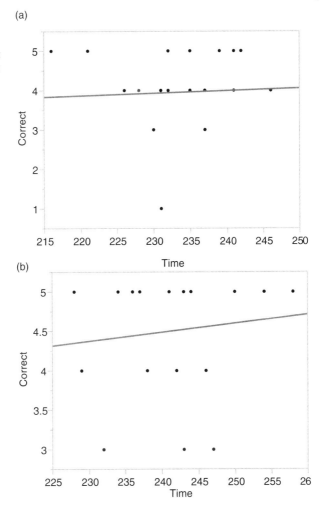

Understand the relationships in the data

It would be easy to run the first test and conclude that the results were statistically significant, and stop. But that leaves the underlying reasons for that significant finding unresolved. The purpose of a research study is not to determine statistical significance, but to explain why the results are significant. [See on page 34.]

By performing the analysis with the data structured in different ways, such as collapsing the data into various binary configurations, you gain a deeper understanding of why the overall result is significant. In this case, we found a relationship between the number of drafts and the perceived assignment difficulty, but then we found no significance when we looked at 1 draft versus 2–4 drafts. This becomes the interesting aspect that makes the research worthwhile. What we need to uncover and discuss in the research report is a more detailed examination that explores the number of drafts versus perceived difficulty and explains how the relationship actually plays out.

Difficulty Rating

The difficulty rating was scored by two different researchers. We need to test how closely they agreed. Since they watched the same interaction, ideally they should have given the same difficulty ranking. However, no group of people will achieve that ideal, so we need to test to see how far from complete agreement they really were. A low inter-rater reliability [defined on page 87] seriously compromises the results of a data analysis.

Ensuring high reliability is a study design issue and requires calibration of the raters.

Cohen's kappa [defined on page 89] is used to compare the agreement between two people— for three or more people, use Fleiss kappa. Kappa values do not return a p-value, but an agreement (kappa) value. Conventionally, a kappa of <0.2 is considered poor agreement, 0.21–0.4 fair, 0.41–0.6 moderate, 0.61–0.8 strong, and more than 0.8 near complete agreement.

The Cohen's kappa for the two raters was very low (Table 7.3).

The Kappa statistic compares an observed accuracy with an expected accuracy. With values of 1–5, even if the raters gave random values, there would be 20% agreement. Cohen's kappa takes that into account to calculate how closely the values agree.

Table 7.3 Cohen's kappa results for reading difficulty for low light.

		Kappa	SE	95% Confidence interval
Low light	Unweighted	0.2857	0.1445	0.0026–0.5689
	Weighted	0.4074	0.1444	0.1244–0.6904
Normal light	Unweighted	0.1875	0.1333	−0.0738–0.4488
	Weighted	0.4624	0.0998	0.2668–0.6579

Calculated from http://www.statstodo.com/CohenKappa_Pgm.php.

The weighted kappa values should be used if the variables have more categories than binary (yes and no).

The 95% confidence interval calculated as kappa $+/-1.96$ SE (standard error).

With a weighted value of 0.40 for the low light and 0.46 for the normal light conditions, the agreement is at the boundary of poor and moderate agreement. The study design should have better controlled for the raters giving values. The poor/moderate value means that any statistical tests the relate the difficulty rating to other study values will need to be qualified in the study report.

t-Test cannot replace doing a Cohen's kappa test

A *t*-test of the raters values are not equivalent to performing a Cohen's kappa test for agreement. From a statistical view, it makes no sense to do a *t*-test of rater 1 versus rater 2 (Figure 7.11). They should be independently giving values that agree.

If a *t*-test is done on the low light condition ratings, they show the two-tailed test is just slightly more than being statistically significant. From this we could conclude that the two raters were different, but it gives nothing about how well the agreed or did not agree on their ratings.

Figure 7.11 *t*-Test for rater agreement. The two-tailed *p*-value does not indicate significance, but the one-tailed does. Either way, a *t*-test should never be used as a method of calculating inter-rater agreement.

t-Test: Two-Sample Assuming Equal Variances

	Rater 1	Rater 2
Mean	2.500	1.800
Variance	1.211	1.326
Observations	20.000	20.000
Pooled Variance	1.268	
Hypothesized Mean Difference	0.000	
df	38.000	
t Stat	1.965	
P(T<=t) one-tail	0.028	
t Critical one-tail	1.686	
P(T<=t) two-tail	0.057	
t Critical two-tail	2.024	

The *p*-value, interpreted with respect to the low Cohen's kappa value, is not surprising. Raters who have a low kappa score would be expected to have sets of scores that would rate low on a *t*-test.

Exercises

Student Questions

1. When we tested the two independent variables for differences between sections, one paragraph said:

    Two-tailed tests are appropriate in both cases since there is no reason to expect either class to be better or worse than the other one. (If the second class had used a different instructional method which we wanted to show had improve their scores, then a one-tailed test would be used.)

    Justify why we should use either a one-tailed or two-tailed test for this study. Should low light reading time or comprehension be assumed to be worse than normal light? Also, explain what would be different in if the study used the other test.

2. A factor in inter-rater reliability is how many levels of ratings (0–1, 1–5, etc.). How does the choice affect both testing for inter-rater reliability and interpreting the results.

3. Figure 7.8 gives results for both a *t*-test and a Wilcoxon test. What are the differences between the tests? Which would be more appropriate for the data in Figure 7.8?

Data analysis

1. The data analysis in the chapter performed the analysis on the low light condition. The same tests need to be done on the normal light condition and the results compared.
    Perform the proper exploratory and inferential statistical tests to determine if there are any correlations for these two relationships for reading order effects for the normal lighting condition.
    You need to determine:
    - Which test to use.

    - Should the one-tailed or two-tailed version be used and why.

    - What is the *p*-value for the test and is it significant?

2. In the exploratory analysis, Figure 7.4 looked at the rating by light level by summing the two raters. The other option would be to average the two raters. Create a graph for the average rating by light level. Compare that

graph with Figure 7.4 and discuss how the differences can affect the interpretation and the exploratory analysis.

3. Create two sets of random data that fits a normal distribution but has unequal variances. Perform a *t*-test that assumes both the equal and unequal variance conditions. Examine how the results vary and how they could affect a study's outcome.

4. Collapsing data into smaller or binary categories changes how the data looks when graphed and, possibly, the analysis results.

    Create data bins for the following:

   a) Create two answer bins: 5 correct and 1–4 correct.

   b) Create five equal-sized bins for the reading times.

      Perform the proper exploratory and inferential statistical tests to determine if there are any correlations for these new relationships for the low and normal lighting condition.

      You need to determine the following:

      • Which test to use.

      • Should the one-tailed or two-tailed version be used and why.

      • What is the *p*-value for the test and is it significant?

Additional Data Sets

   Action-score. Data in: Chapter 7—action-score.xlsx

8

Analysis of Usability of an E-Commerce Site

Usability of an E-Commerce Site

This example, which looks at the data collected during a usability study, covers the typical analysis process for any set of data where the statistical analysis goal is to find statistical significance [defined on page 19]. Note, however, that although the statistical analysis goal is finding significance, that simply finding significance is never a study goal. The study goals must relate back to the context of the study, and the statistical analysis provides element in making the study claims [see page 34].

Study Overview

A study looked at three different designs of an e-commerce site that sold a wide variety of low-priced items. The site design team wants to know which design will produce the highest sales and lowest number of customer interaction problems.

The study used a between-subjects design [defined on page 199] because of the following:

- All three designs had the same items for sale.
- Overlaps in the basic designs and checkout procedures would be expected to show order effects if a within-subjects study design was used.

Subjects were asked to buy a few items for a friend's upcoming birthday. They were provided with a credit card number and told to ship the items to their own home (not the friends).

Introduction to Quantitative Data Analysis in the Behavioral and Social Sciences,
First Edition. Michael J. Albers.
© 2017 John Wiley & Sons, Inc. Published 2017 by John Wiley & Sons, Inc.
Companion website: www.wiley.com/go/albers/quantitativedataanalysis

Data collected was

design   The site designs the subject used. Values are 1, 2, and 3.
total sale   Total amount of money that the person spent on the
                        site.
usability   The usability rating the subjects gave the site after
                        completing the purchase. A Likert scale was used: 1 =
                        very usable, 2 = usable, 3 = neutral, 4 = unusable, 5 =
                        very unusable.
items purchased   Total number of items that were purchased.
success   The success of making the purchases. The researcher
                        recorded whether or not the subject had successfully
                        completed the purchase process. This is independent of
                        whether the subject believed they were successful. Val-
                        ues were 0 = unsuccessful and 1= successful.
age   Age of the subject.
experience   Self-reported experience with purchasing items online.
                        A Likert scale was used: 0 = less than once a quarter,
                        1 = less than once a month, 2 = couple times a month,
                        3 = at least weekly, 4 = every few days.

A power test [defined on page 31] should be performed early during the study design. However, a power test requires the mean and standard deviation values for the three groups. None of the data collected for this study would show a normal curve—except for total sale, everything collected in this study is ordinal data—so it cannot even be used to do an after data collection check. The 25 subjects in each group will hopefully provide adequate power. Clearly, 25 subjects are better than a smaller number, such as 5–7, which would have minimal predictive power.

Know How You Will Analyze the Data Before Starting the Study

Units of Analysis

The main unit of analysis [defined on page 108] for this study is the design level: the basic question to answer is "which design is better?" Any study needs to define its principle level of analysis early in the study development. Since this study wants to determine which design is best, then that defines the main unit of analysis.

Some follow-up analysis will use a unit of analysis of individual people. These follow-up studies will occur after determining the most usable design. Individual people do not make sense as a main unit of analysis, since the study is concerned with testing a design's usability. Tests with individual people involve tests such as dividing the people by age and testing for differences between groups. After determining the design rankings, follow-up tests[defined on page 94] using age within a design could show that younger or older people find different designs more usable. A finding that success varies with age can have profound practical implications [see page 34]. A redesign may cause problems for one group that more than offset any gains by the other group. Younger people might be able to order faster, but increased tech support costs for old people would result in more, not less, expense for the redesign.

Relationships to Explore

Deciding what data relationships to examine is independent of the tests [defined on page 54] used to examine them. Ideally, they should be determined early in the study design and before considering which tests to use. The data relationships of interest come directly from the study hypothesis, or more precisely, the intent/reason for the study hypothesis [defined on page 15]. Written as a null hypothesis, it may say two variables will not affect each other. In reality, the study hopes to show how they affect each other. The statistical analysis supports the claim.

The following are the main relationships of interest:

- **Design–Total Sale.** Determines if the design affects the total sales on the site. It would be nonsignificant if people spent similar amounts on each site. A significant finding provides evidence for which site is better.
- **Design–Items Purchased.** Determines if the design affects the number of items that are purchased. It would be nonsignificant if people made similar number of purchases on each site. A significant finding provides evidence for which site is better.
- **Design–Usability.** Determines if the design affects how usable people find a site. If all of the designs are similar, it would be nonsignificant. In this study would provide a basis for claiming one site design is better than the other two.
- **Design–Success.** Determines if the design affects how people were able to successfully complete a purchase. If all of the designs are similar, it would be nonsignificant. In this study, we want to find significance to have a basis for claiming one site design is better than the other two.

Follow-up relationships within a specific design:

- **Usability–Success.** Compares people's self-reported feelings of specific design's overall usability to their actual success. They should be highly correlated.
- **Success–Experience.** Determines if successfully completing a purchase is related to a person's experience with online purchases. Should be non-significant since any user should be able to successfully purchase items.

- **Usability–Age.** Determines if any age group finds a specific design more or less usable. Should be nonsignificant.
- **Success–Age.** Determines if any age group has more or less trouble successfully completing a purchase with a specific design. Should be nonsignificant.

The two major factors that the study needs to examine are the design–usability and design–success relationships. These two will form the basis for ranking the overall quality of the three designs. The usability–success relationship then determines if those two factors are correlated for a design.

Deep analysis, not a superficial analysis

Consistent finding across many usability studies is people will report sites as highly usable even when they experience major problems and do not successfully complete the task (Spool, 2005). Similarly, in surveys, what people say they do often conflicts with what they actually do. This quirk of human nature is well supported in the psychology and sociology literature, but many study's results fail to take it into account. Failing to realize the difference can seriously skew a study's results.

Part of the reason for performing a deep analysis of the data is to uncover these types of contradictions and take them into consideration when reporting the practical results of the study. Note how a superficial analysis would simply find and report the significance of each factor without considering how they relate to each other. It is important to verify the results of different tests make sense and are not in conflict, such as people like site 2 the best but perform poorly on it.

Perform an Exploratory Data Analysis

An exploratory data analysis [defined on page 108] is the time to examine the data and develop a feel for how the numbers relate to each other. This gives a basis on judging both the results of the inferential statistical tests and how those results connect back to the study's hypotheses. It may also result in seeing potential or interesting connections that deserve further analysis but were not part of the original data analysis plan. If you jump directly into inferential statistics, you risk finding statistical significance but not having a deep enough understanding of the data to determine:

- If the statistical significance has any practical implications [defined on page 34].

- If the statistical significance of one data element conflicts with a statistically significant finding of another data element.
- What deeper analysis should be explored to expand upon and more fully understand the initial iteration of the data analysis[defined on page 6].

Descriptive Statistics

An exploratory data analysis typically starts with descriptive statistics (Table 8.1).

The gray areas in Table 8.1 are data that has little meaning. See the sidebar "Descriptive statistics of with nominal and ordinal data" for an explanation.

Table 8.1 Descriptive statistics for the three designs.

	Design	Median	Mean	Standard deviation	Skewness	Kurtosis
Total price	1	9.87	11.30	5.03	2.55	7.65
	2	13.53	13.36	5.36	0.21	−1.13
	3	18.12	17.00	5.90	0.36	−0.05
Usability	1	2.00	1.84	0.99	0.91	−0.21
	2	2.00	2.60	1.26	0.44	−0.86
	3	4.00	3.56	1.26	−0.69	−0.53
Items purchased	1	2.00	1.60	0.65	0.61	−0.48
	2	1.00	1.64	0.76	0.73	−0.81
	3	3.00	2.36	0.91	−0.82	−1.31
Success	1	0.00	0.32	0.48	0.82	−1.45
	2	0.00	0.24	0.44	1.30	−0.35
	3	1.00	0.68	0.48	−0.82	−1.45
Age	1	34.00	35.16	5.14	−0.39	0.71
	2	35.00	34.8	4.64	−0.68	−0.04
	3	35.00	35.12	4.97	0.00	−1.29
Experience	1	2.00	2.00	1.50	−0.08	−1.45
	2	2.00	2.12	1.33	−0.12	−0.87
	3	2.00	1.84	1.46	0.04	−1.34

A visual inspection can reveal if any of the numbers seem to be different from the others. The following can be seen in this table:

Total sale | Comparable means for design 1 and 2, but design 3 looks different. Something in the design was driving these changes and it requires an explanation in the final report. Also, since the median [defined on page 37] and mean are similar, the total sale may be normally distributed.

All of the standard deviations [defined on page 39] are similar, which is a good thing. It means the variance of total sales is similar in all three cases—in other words, all three total sale curves should look similar. The standard deviation is related to the variance and t-tests and ANOVA work best when they have an assumption of equal variances. For design 1, skewness [defined on page 66] and kurtosis [defined on page 69] show the curve is very skewed and much higher peaked than would be expected from a normal curve (normal curve kurtosis $= 0$). Design 2 and 3 both have relatively low values that are compatible with a normal distribution.

Usability | Comparable for design 1 and 2, but design 3 looks different. Since this was scored with a Likert scale of 1–5, the use of both median and mean values have to be used carefully.

Items purchased | Comparable means and standard deviations for design 1 and 2, but design 3 looks different. The medians are different for all three designs. Skewness and kurtosis show the curves should be close to symmetrical.

Since items purchased has a spread of only has a few small numbers, it is easy to read too much into these values, but they do deserve graphing and a follow-up examination.

Success | Design 1 and 2 have a median of 0 and design 3 has a median of 1. Success was scored as a binary; the mean, standard deviation, skewness, and kurtosis values are meaningless.

Age | Comparable across all three designs. This is good since age is not a study design factor and it means that age-related factors probably did not confound the study.

Skewness and kurtosis show the curves should be close to symmetrical, although for design 3 with a kurtosis $= -1.29$, the curve will be somewhat flatter than a normal curve. In a random sample that did not consider age, age should be a uniform distribution with equal numbers of each age.

Experience   Comparable across all three designs. This is good since
                              experience is not a study design factor and it means that
                              experience-related factors probably did not confound
                              the study.

Taken as a whole, the inspection of the descriptive statistics shows that in some factors all three designs are comparable and in others design 3 stands out. Design 3 has more favorable values on all of the values we expect to change if the designs are different. As we continue with the analysis, we should expect design 1 and 2 to have nonsignificant differences and for design 3 to have significance differences. As the researcher, you would then be expected to go back to the designs and figure out what features caused those differences.

The similar descriptive statistics of all three design categories for age and experience are a good thing. It means that the subjects were divided into similar groups. Any group of people will show variation, the similar standard deviations mean we can be confident that all of the groups had similar variations. This helps to show there was no biasing [defined on page 208] when group assignments were made and that each group reflects the overall user population, assuming the overall sample reflects the population. If one or more groups do not have similar variations, the data analysis becomes more complicated and loses power. If the three groups had substantially different standard deviations, it would be prudent to reexamine the subject selection methods to rule out any selection bias. It would also be prudent to consult with a statistician. Yes, the designs may naturally have very different standard deviations, but part of the data analysis is showing those differences arise from the design itself.

Descriptive statistics of with nominal and ordinal data

The descriptive statistics for the variables with nominal [defined on page 302] and ordinal [defined on page 302] data (usability, success, and experience) must be examined with the understanding that mean and standard deviation have little meaning for this type of data.

Clearly, for nominal data, where the values could be in any order all of the descriptive statistics are meaningless. Consider a data set where different departments were coded 1 = accounting, 2 = receiving, 3 = production, etc. A mean of 2.4 tells us nothing since the number to department assignments are arbitrary.

For ordinal data, the differences between individual values are different (or more/less undefined). Usability is scored 1–5, but the difference between strongly agree–agree (4–5) is not the same as agree–neutral (4–3). As a result, the average can only be taken a general indicator of how the overall group rated the usability and not as a value to report as "the group average and standard deviation." Unfortunately, reporting "group average and standard deviation" is a very common occurrence without any qualifying statements.

Graphical Exploratory Analysis

To get a better feel for the data, either scatterplots [defined on page 80] or histograms [defined on page 79] (depending on the data type) need to be created. This will reveal the overall shape of the distributions and any outliers. The descriptive statistics led us to believe design 1 and 2 will be similar and design 3 will be different. Graphing the data allows us to visually verify that and to see if the difference is enough that we expect it to be statistically significant. When we look at the graphs, we want to see if the three designs are generally matching or if they deviate from each other. We also want to check between graphs to see if they show a consistent trend. For example, usability and success should both be higher on the better design; if one increased and the other decreased, we would have to probe deeper to determine the cause.

The data for total price and for age are binned into increments of 5—either $5 or 5 years—to make the trends clearer. For example, all values from $15 to $19.99 appear in a single bar and are treated as a single group. The reason to bin data is because otherwise a histogram would be flat with each value occurring only once or twice. If a study has several hundred data points, then unbinned data would provide a good histogram.

Total sale | Figure 8.1. All three look very different. Designs 1 and 2 have much lower total amount purchased than design 3, thus something about design 3 must have been encouraging people to either buy more items or buy higher priced items. The statistical data will not be able to resolve this question; the researcher needs to go back to the test notes or videos to figure out what features caused this difference.

Figure 8.2 shows the same data, but without the data bins. The clustering of the data is less distinguishable since there are so many values that only occurred once or twice.

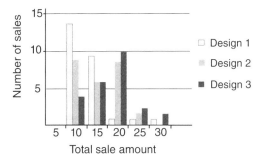

Figure 8.1 Comparison of total sale. Bin size was in $5 increments. It is clear that design 1 and 2 had lower total price purchases than design 3. The values were placed in $5 bins because otherwise the graph would be flat with each price occurring only once or twice.

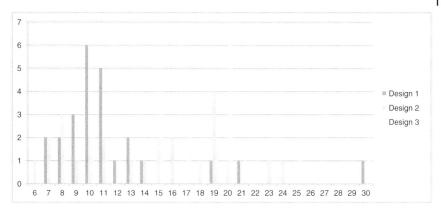

Figure 8.2 Comparison of total sale with raw data.

This graph helps explain the different skewness [defined on page 66] and kurtosis [defined on page 69] values that were found for design 1. Notice that design 1 is highly skewed to the left with a very long flat tail. It is not even remotely a bell-shaped curve and that is reflected in the skewness and kurtosis values.

Usability   Figure 8.3. All three look very different, with design 1 skewed left, design 2 flat, and design 3 skewed right. The possible reasons for this will need to be examined closer in the later analysis.

Looking back at Table 8.1, the means and median for designs 1 and 2 are comparable. Similar means and median values do not necessarily imply a similar overall distribution. The very different visual appearance in the graph shows why data needs to be viewed in multiple ways. It also shows a drawback of trying use mean/median with ordinal data.

Figure 8.3 Usability rating given by the subjects for each design.

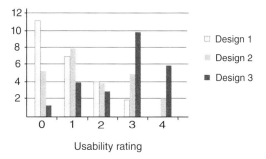

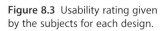

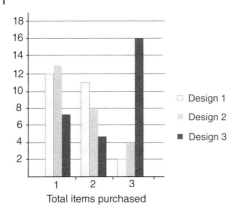

Figure 8.4 Items purchased for each design.

Total items purchased

Looking at the total price and usability together, it is becoming apparent than design 3 seems to the best design, at least it received the highest usability ratings and highest sales.

Items purchased | With designs 1 and 2, most people purchased only 1 or 2 items with 1 items dominating (Figure 8.4). With design 3, several people purchased 3 items, which only occurred in a few cases for designs 1 and 2.

Because people purchased more items with design 3, it seems reasonable that their total purchase price would be higher, which can be seen in Figure 8.4.

Success | Since success was evaluated as a pass/fail, all three designs were compared on a single- or multiple-bar graph (Figure 8.5). Design 3 seems to be much more success for than design 1 and design 2. Remember this is

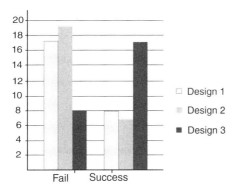

Figure 8.5 Success for the three designs.

Fail   Success

(a)   (b)

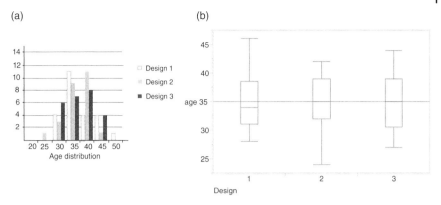

Figure 8.6 Age range of the subjects, compared with both a column graph and a box plot.

a binary [defined on page 77] and researcher-scored valued.

The number of failures on all three designs seems to be overly high. It would be worth looking at the failure rates for 2 or 3 items purchased versus 1 item purchased to determine if multiple items are the cause of the failure.

Age

The ages show a wide spread, but none of three graphs reveal any clustering or distribution that does not compare to the other two (Figure 8.6).

Looking at and comparing different plots of the same data can help expose different aspects and helps the researcher get a deeper feeling for the data. The column graph gives a good feel for how the groups match up against each other. It is clear that each age bin has a similar number of subjects. We can also see that design 2 subjects are tightly clustered at 35 and 40 age groups. The box plots give a good visual of the spread of the data. The boxes—with 50% of the data—are all similar in size and placement. The tails for design 2 are lower at both the top and bottom, but not enough to be of concern.

Experience

The experience shows a rather equal distribution across all of the possible values (Figure 8.7). As an independent variable in the study, this is a good thing for the analysis since we can compare various levels of experience to see if it changes the *Usability* or *Success* factors.

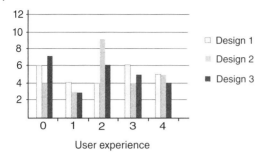

Figure 8.7 Range of subject's experience in online purchases.

Examining the different graphs and comparing them allows a researcher to start to draw conclusions and to see if the study's hypotheses are being supported. However, just examining a graph does not reveal if the changes are significant or if they are just random variation; answering this question requires performing a statistical test. With small data sets, random variation can make it appear graphically as if the different designs are actually different when they are not. In turn, this can lead to design decisions being made that can be costly to implement with little return on the investment [explained on page 24].

Comparing graphs requires the same *x* and *y* axis values

Comparing data visually, as done with Figures 8.1–8.7, requires taking care the graphs used are effective for the comparison and that they are created with the same scales. The choice of how to plot the data (bars, lines, etc.) depends on the data itself, but it must make sense for representing that data. If multiple graphs are used, they must all have the same *x*- and *y*-axis values. Excel and other graphics programs will autoadjust the scales to values that maximize the data spread for the individual data, but this can severely impact visual comparison. You will need to adjust the *y*-axis values, so all graphs have the same data range.

A brief comparison of these two graphs could conclude that the graph on the left has higher values (Figure 8.8). But a close look shows the *y*-axis scales are completely different. The data presentation problem is that people will compare graphs that are close together, even if the data itself do not support the comparison. Plus, many decisions are made long after looking at the data; in 4 days, a person only has a mental image of two similar graphs, not that the scales were different.

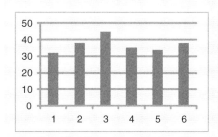 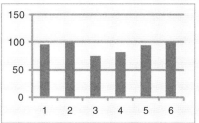

Figure 8.8 Comparison with unequal scales. At a quick glance, the two graphs look similar, but they have very different scales.

Perform an Inferential Statistical Analysis

Having completed the exploratory analysis, we expect to find design 3 to be different from the other two designs. Now we need to perform an inferential statistical analysis to verify that our initial and visual conclusions are actually correct.

First, we need to verify the following:

- Data supports the assumptions of the statistical tests. With three designs, an ANOVA, which assumes the data has a normal distribution, would be used.
- The subjects groups were not biased in some way. With smaller samples, even random assignments may result in the groups not having a similar make up. In this case, age and experience were the two factors that should be the same across all the groups. The data analysis must verify they do not vary between groups.

Verify Total Sale and Age Show a Normal Distribution

ANOVA assumes a normal distribution of the data. Thus, we should test to ensure the interval data (total sale and age) both show normality [defined on page 64]. Total sale and age are the only continuous variables [defined on page 201] in the study; thus, they are the only ones where a normality test is appropriate [defined on page 66].

We will use the Shapiro–Wilk normality test. As with any statistical test, we can define the acceptable p-value. In this case, we will use $p = 0.1$, a less strict value than the $p = 0.05$ that we use for the ANOVA test. However, the ANOVA is somewhat forgiving of data that close to normal and the Shapiro–Wilk normality test needs more than 25 data points for a high confidence. An online version is available at http://dittami.gmxhome.de/shapiro/

Table 8.2 Total sale: Shapiro–Wilk normality test.

Design	N	Mean	SD	W	Threshold	Hypothesis
1	25	11.30	5.03	0.72	0.93	Reject
2	25	13.36	5.35	0.94	0.93	Fail to reject
3	25	17.00	5.90	0.95	0.93	Fail to reject

Table 8.3 Age: Shapiro–Wilk normality test.

Design	N	Mean	SD	W	Threshold	Hypothesis
1	25	35.16	5.13	0.92	0.93	Fail to reject
2	25	34.80	4.63	0.95	0.93	Fail to reject
3	25	35.12	4.97	0.95	0.93	Fail to reject

Each of the three distributions is analyzed separately with the Shapiro–Wilk normality test. The test returns a W value, which is comparable to a p-value. It also calculates the threshold values that W must exceed in order for the test to be nonsignificant. With this test, a significant W value means the data is not normally distributed. In this case, we want nonsignificance since ANOVA wants normal data and rejecting the hypothesis would mean the data is not normal.

The test shows Design 1—*Total sale* is not normally distributed and designs 2 and 3 are normally distributed (Table 8.2). Looking at the graphs for total purchase (Figure 8.4), we can see that they support this conclusion. Design 1 is highly skewed to the left. Notice how design 2 is rather flat yet this test says it is normal. Likewise, Table 8.3 shows the results for age, with all three designs having normally distributed values. Looking at the graphs for age (Figure 8.6), the distribution appears skewed.

Normality test have low power with a small number of data points

The 25 data points we have are more or less the minimum for acceptable normality test results. With the sample sizes typical in social science research, it is rare to have a sample large enough to give a normality test much power. On the other hand, the low power with small numbers means that when the test does give a result of nonnormality, you can assume this is a real result and must take it into consideration when performing the statistical tests.

The low power with small numbers may be a reason why normality tests are often skipped in social science research.

Verify Age and E-Commerce Experience Are Not Significant

As a general rule, you do not want the demographic variables to be significant between groups unless you are defining them that way for the study (e.g., a comparison of 20–29 versus 50–59 year old groups).

The exploratory analysis leads us to believe that both age and experience will not be significant between the three subject groups, which is the desired result. The subjects were randomly distributed, so both age and experience should also be randomly distributed across the groups. If they were randomly distributed, then the statistical tests would show the difference between the three groups to be highly nonsignificant.

Age

Age is an interval variable and there are three groups, so an ANOVA is an appropriate test. We have already verified the data fit a normal distribution. When the test is run this time, it is only looking at a single factor (age) across all three groups, so the ANOVA Single Factor test should be used [defined on page 94]. The results are shown in Table 8.4.

Table 8.4 Statistical calculation for the age variable.

ANOVA

Summary

Groups	Count	Sum	Average	Variance
1	25	879	35.16	26.39
2	25	870	34.80	21.50
3	25	878	35.12	24.69

Source of Variation	SS	df	MS	F	p-Value	F Crit
Between groups	1.95	2	0.97	0.04	0.961	3.12
Within groups	1742	72	24.19			
Total	1743.95	74				

Levenes test for homogeneity of variance

	df	F Value	Pr($>F$)
Group	2	0.506	0.605
	72		

The variances between the groups are similar and Levene's test—for equal variances—has a p-value of 0.605, which supports that the variances are the same [defined on page 39]. Thus, we do not have to worry about using the ANOVA for unequal variance. Notice that in Table 8.4, the variances have values ranging from 21 to 26, but without a formal test, we do not know if this is random variation or a true difference. The data need to be formally tested to resolve that question; a researcher cannot resolve it by just looking at the data. If they were extremely different, more advanced statistical methods may be appropriate; consult with a statistician.

For the ANOVA, the p-value is 0.961 [defined on page 19], which means the age differences are insignificant, the desired result.

Experience

Experience was ordinal Likert scale data, thus an ANOVA is not appropriate. Instead, the nonparametric Kruskal–Wallis test was performed (Table 8.5) [defined on page 95]. The results had p-value of 0.811, so we know the differences in experience across all three were not significant, which is our desired result.

Results of a Kruskal–Wallis test from http://vassarstats.net/kw3.html

As a comparison, an ANOVA run on *Experience* also shows a nonsignificant result (Table 8.6). In this case, the Kruskal–Wallis and the ANOVA have similar p-values, but this is happenstance. Comparing p-values across tests is meaningless.

Both p-values here are about 0.8, which is highly nonsignificant. On the other hand, sometimes the ANOVA test may give a statistically significant result and the Kruskal–Wallis test will not (or vice versa). There is typically one best test to perform and the analysis needs to use that one. It is poor data analysis methodology to run both and select the one that gives a significant result.

Table 8.5 Kruskal–Wallis test results for experience.

Mean ranks for sample

Design 1	Design 2	Design 3
38.3	39.8	35.9

$H =$ 0.42
df $=$ 2
$p =$ 0.8106

Table 8.6 ANOVA results for experience.

ANOVA: Single factor

  Summary

Groups	Count	Sum	Average	Variance
1.00	25.00	50.00	2.00	2.25
2	25.00	53.00	2.12	1.78
3	25.00	46.00	1.84	2.14

ANOVA

Source of Variation	SS	df	MS	F	p-Value	F crit
Between groups	0.99	2.00	0.49	0.24	0.79	3.12
Within groups	148.00	72.00	2.06			
Total	148.99	74.00				

Analysis of Variance

Source	DF	Sum of Squares	Mean Square	F Ratio	Prob > F
Design	2	416.6293	208.315	7.0362	0.0016*
Error	72	2131.6497	29.606		
C. Total	74	2548.2790			

Figure 8.9 ANOVA results for total sales.

Total Sales

Figure 8.9 shows the graph of the total sales by design. It seems that people using design 1 spent less than people using the other two designs did.

Because the total sale price is a continuous value, we can use an ANOVA to analyze the data. We find that the differences are highly significant.

Since the ANOVA assumes the data has equal variances [defined on page 94], we need to perform a Levene's test to check that assumption (Figure 8.10). It returns a $p = 0.1999$, so the assumption of equal variances is supported. JMP actually performs several different tests for equal variances. We can see all of them reject the hypothesis, as desired.

Follow-Up Tests

The ANOVA result of statistical significance [defined on page 19] means that at least one of the pairings (1–2, 1–3, 2–3) are statistically different, but it does not

Level	Count	Std Dev	MeanAbsDif to Mean	MeanAbsDif to Median
1	25	5.034612	3.149504	2.839600
2	25	5.350793	4.603808	4.597200
3	25	5.902579	4.671488	4.626800

Test	F Ratio	DFNum	DFDen	Prob > F
O'Brien[.5]	0.2024	2	72	0.8172
Brown-Forsythe	1.9852	2	72	0.1448
Levene	1.6466	2	72	0.1999
Bartlett	0.3065	2	.	0.7360

Figure 8.10 Test for equal variances for total sales.

indicate which one. Thus, we need to perform follow-up tests [defined on page 94] to determine which of the designs are different. Follow-up tests are only appropriate if the initial test (in this case the ANOVA) indicates significance; do not skip the initial test and jump directly to the follow-up tests. This greatly increases the risk of a false positive.

This example was done with SPSS and uses a Tukey HSD test. t-Tests for all comparisons (1–2, 1–3, 2–3) could also be performed.

The Tukey HSD follow-up tests (Table 8.7) show that designs 1 and 2 do not have a significant different, but that design 3 does show significant differences from designs 1 and 2. If we compare this result to the graphs in Figure 8.1, we can see that although design 3 looks different, there also visually appears to be a

Table 8.7 Results of the Tukey HSD test.

Tukey HSD

(I) Design	(J) Design	Mean difference (I – J)	Std. error	Sig.	95% Confidence interval	
					Lower bound	Upper bound
1	2	−2.06320	1.53899	0.378	−5.7462	1.6198
	3	−5.70120*	1.53899	0.001	−9.3842	−2.0182
2	1	2.06320	1.53899	0.378	−1.6198	5.7462
	3	−3.63800	1.53899	0.054	−7.3210	0.0450
3	1	5.70120*	1.53899	0.001	2.0182	9.3842
	2	3.63800	1.53899	0.054	−.0450	7.3210

The Sig column is the p-values. Notice in the 95% confidence interval column that the range for the nonsignificance findings includes 0.

difference between design 1 and 2, which is not supported by the statistical analysis [discussed on page 15]. The statistical significance of design 2 and 3 can argued; rounding gives $p = 0.05$. Once again, we have a reminder that statistical significance cannot be determined by simply looking at a graph.

Items Purchased

Figure 8.4 shows the graph of items purchased. It appears that with designs 1 and 2, most people purchased only one or two items, while with design 3 most people purchased three items. Thus, we expect to see a significant difference.

The items purchases are obviously an integer number greater than 0. With continuous numbers, a single-factor ANOVA would be the appropriate test. An argument can be made to use a nonparametric test because there are only three values in each group and the distribution is not normal. We will run both the ANOVA and the Kruskal–Wallis test to see if there is any difference in the results (Table 8.8).

Both tests show that there is a significant difference between the three designs. For the ANOVA, $p = 0.001$ and for the Kruskal–Wallis test $p = 0.006$.

Performing Follow-Up Tests

Since the ANOVA showed a significant difference, we need to perform follow-up tests to determine which of the designs are different [defined on page 94]. We could also use the nonparametric Mann–Whitney test as a follow-up for the Kruskal–Wallis. The nonparametric tests would show the same results for significance.

This example uses SPSS and does a Tukey HSD test. t-Tests could also be performed.

The Tukey HSD follow-up tests (Table 8.9) show that designs 1 and 2 do not have a significant different, but that design 3 does show significant differences from designs 1 and 2. The nonparametric follow-up test is presented as student exercise #2.

Success

Figure 8.11 shows the graph for the fail and success of the three designs. We can see that designs 1 and 2 had high failure rates and low success rates, while design 3 had the opposite. Thus, we expect to find a significant difference.

Table 8.8 Test results for items purchased.

A

ANOVA: Single factor

Summary

Groups	Count	Sum	Average	Variance
1.000	25	40	1.600	0.417
2	25	41	1.640	0.573
3	25	59	2.360	0.823

ANOVA

Source of Variation	SS	df	MS	F	p-Value	F crit
Between groups	9.147	2	4.573	7.566	0.001	3.124
Within groups	43.520	72	0.604			
Total	52.667	74				

B

Kruskal–Wallis test

Results of a Kruskal-Wallis Test from http://vassarstats.net/kw3.html

Mean ranks for sample

Design 1	Design 2	Design 3
32	32.7	49.3

$H =$ 10.16

df = 2

$p =$ 0.0062

A. ANOVA results. B. Kruskal–Wallis results.

The success variable is a binary value; it is either 0 or 1, depending on if the researcher determined the person successfully completed purchasing the items. This means that we need to use a test appropriate for binary data [defined on page 95]. Unlike our previous examples, neither the ANOVA nor the Kruskal–Wallis tests are appropriate here. Instead, we need to use a chi-squared test.

JMP provides a visual of the three options and performs both a chi-squared test and a Pearson test. Both are showing the differences are highly significant.

Table 8.9 Follow up test results for items purchased.

Tukey HSD

(*I*) Design	(*J*) Design	Mean difference (*I* – *J*)	Std. error	Sig.	95% Confidence interval Lower bound	Upper bound
1	2	−0.040	0.220	0.982	−0.57	0.49
	3	−0.760*	0.220	0.003	−1.29	−0.23
2	1	0.040	0.220	0.982	−0.49	0.57
	3	−0.720*	0.220	0.005	−1.25	−0.19
3	1	0.760*	0.220	0.003	0.23	1.29
	2	0.720*	0.220	0.005	0.19	1.25

Tukey HSD results for items purchased analysis. Design 3 is significantly different from design 1 and 2.

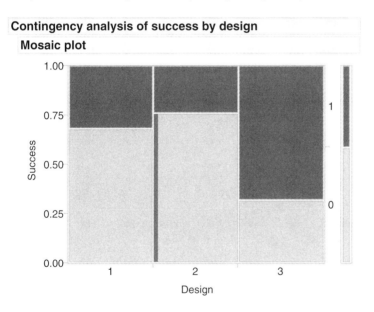

Figure 8.11 Chi-squared test for success. The differences are significant.

Statistical test should be triangulated against any other available data

The results of any statistical test should be triangulated against any other available data. The data in this example showed significant findings for total sales and items purchased. From this, we know the differences have a low probability of being caused by random variation in the data. More formally, that the test cannot support a claim that the two sets of data come from different populations. But the study report should explain what is causing those significant findings, something that a statistical test alone cannot do.

Other sources of data (often qualitative) may be able to explain the cause of the significance. In this usability study, recordings of subjects interacting with the information should be available. How the specific causes are found can vary and ways of triangulating can vary, for example:

- A review of the video can show the specific points where subjects had difficulties with interacting with the options.
- A post hoc review of the session with the subjects would provide a transcript of why they performed specific actions.

The triangulation goal is to correlate the statistically significance findings and to relate them to how people were seen interacting with the information.

Of course, when doing a triangulation, you must use different information. If the video recording had been coded and the statistical tests performed on that coding, then obviously the video would match the test results.

Two Factor Number Crunching

Up to this point, we have been looking at the design versus a single factor. The next step up is to look at two factors together and if/how they affect a third factor [defined on page 50]. This next level of statistical testing looks at the interaction between factors. For example, do total sale and usability interact to affect success? Or do they interact to affect the number of items purchased. The value in looking at two factors is that a significant finding could reveal that although neither factor alone (such as total sale and usability) show a significant result, together they interact and significantly affect the third factor (success). Two-factor analysis may look at the data to determine the following:

- Total sale × usability against success
- Total sale × usability against total items purchases
- Total sale × total items purchases against usability

It is important to pick the factors, so they are meaningful and not simply test all combinations within the data. The combinations used in the test should have been decided upon during the study design, before any data was collected.

There are multiple statistical tests that can be used (typically a two-way ANOVA). If the study has lots of variables, three-way or four-way ANOVA could also be done. But these tests are complex enough that the use of a statistician is recommended. A major difficulty is setting up the data set to ensure a three-way or four-way ANOVA test is actually testing the desired relationships. In addition, performing these tests requires dedicated statistical software, such as SPSS or JMP.

Exercises

Data in: Chapter 8—usability of e-commerce.xlsx

Student Questions

Define the following terms and explain their importance in data analysis:

1) Follow up test
2) Mean of ordinal data
3) Meaning of a significant result with an ANOVA
4) Normality tests
5) Two-way (or more) interaction between variables

Data Analysis

1. Usability
    Figure 8.3 shows the graphs of the usability variable. We can see that design 1 has low value for usability, design 2 is in the middle, and design 3 shows higher values. Now, we need to test to determine if these differences are significant.

    Perform the proper test for the usability data and interpret the results. Should it be an ANOVA or a Kruskal–Wallis test? Which follow-up tests should be performed if the results show significance?

2. Items purchased
    The Kruskal–Wallis test shows there were significant differences between the groups. Perform the proper nonparametric follow-up tests to determine which designs differ.

Additional Data Sets

Usability test analysis. Data in: Chapter 8—usability test analysis.xlsx

Reference

Spool, J. (2005) Seven common usability testing mistakes. Available at https://articles.uie.com/usability_testing_mistakes/ (accessed June 15, 2016).

9

Analysis of Essay Grading

Analysis of Essay Grading

Some studies focus on building a model (an equation) that describes the relationships between the variables [defined on page 92]. Often this model is more valuable than simply knowing that variables are related in a statistically significant way because it gives how they are correlated [defined on page 55]. For example, with a model, we can know that the dependent variable will increase 1.4 times the increase in the independent variable. This example considers how to analyze the data from a study with a goal of creating the model.

This example looks at the following:

- Inter-rater reliability of the evaluation of student papers. Any time multiple researchers rank or score an artifact, the inter-rater reliability should be calculated to show how closely they agreed [defined on page 87].
- Create data bins and consider how different levels of binning can change the analysis.
- Developing a regression equation that can be used make predict a student's grade. Regression equations are useful in any study that collects multiple variables and wants to use them to predict a variable.

Study Overview

As part of an analysis of the writing abilities of a class, a set of essays written by the students in three different sections (sections 103, 142, and 164) of the same course were read and graded by three different people. Their evaluation consisted of assigning a numerical grade to the paper and rating it on four categories.

Introduction to Quantitative Data Analysis in the Behavioral and Social Sciences, First Edition. Michael J. Albers.
© 2017 John Wiley & Sons, Inc. Published 2017 by John Wiley & Sons, Inc.
Companion website: www.wiley.com/go/albers/quantitativedataanalysis

One goal of the study is to see if the grade can be predicted by paper features. The evaluators assigned Likert rankings of the paper's quality (1 = low 5 = high) in each of four categories.

Grammar   Use of proper grammar and punctuation.
Argument   How well the argument is developed.
Paragraph   Use effective paragraph structure and transitions.
Sentence   Use of varied sentence length and complex structures.

To prevent preferential treatment during the grading, none of the three graders had taught the sections being studied, they did not know what section a paper came from, and the student names had been removed from the papers.

As a calibration procedure for inter-rater reliability, the graders had practiced scoring papers and discussed the reasons for their scoring until they were within an acceptable agreement range on the grades and four categories. This would be tested with the Fleiss' kappa scores discussed below [defined on page 90]. A test value should be established beforehand as the desired goal (a kappa of 0.65 fits most study requirements) and the training continued until that level of agreement between raters is achieved. The expected kappa value needs to be established as part of the study's initial methodology design.

The data analysis will only examine and analyze the grades for the papers as a group. In a more complete analysis, each of the four categories (grammar, argument, paragraph, sentence) judged by the raters should also be examined for both the inter-rater reliability and across sections to verify they are all similar. The procedure for this analysis would be identical to what is described here.

Exploratory Data Analysis

The data generated during this study do not lend itself to descriptive statistics because, except for the grade, they are all ordinal data. For example, grammar is scored 1–5, but the difference between 1 and 2 is not the same as 3 and 4. As a result, an average can only be taken as a general indicator of how the overall group fared on grammar and not as a value to report as "the group average." (Reporting averages of ordinal data is common in research reports, but the values are rarely qualified as a general indicator.)

Compare the Three Section's Grades

The first step in the exploratory analysis is to determine if the three sections have similar scores [defined on page 108]. Whether they should be similar

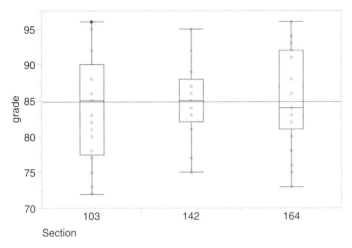

Figure 9.1 Grades by section.

depends on the overall goal of the study. For example, a study evaluating different teaching methods would want to find differences, but a study defining overall student baseline ability would want all sections to be similar. Part of the exploratory data analysis is to gain a feeling if there are expected differences between the sections before continuing. If they are similar, then the values can be combined and analyzed as a single group and, if not, then they will need to be analyzed as three different groups.

Figure 9.1 shows a box plot[defined on page 81] of the grades by section. Section 142 has a smaller midrange of values, but the overall spread of scores is similar. We also see the median scores are close for all three sections. Section 142 can be expected to have a smaller standard deviation, based on the smaller size of the 25 and 75 percentile lines.

Compare the Rater's Scoring

When two or more people are scoring items within a study, it is important that they assign similar scores. Note that the study had the three raters practice scoring papers until they were assigning similar values. The second step would be to look at how the raters performed when scoring the papers in the study. Figure 9.2 shows that rater 3 gave fewer high scores, but that the midrange and low end were comparable. We can also see that rater 2 had the narrowest midrange, which means the scores were more tightly clustered around the median. Later, we will be performing a statistical test to determine if the three raters were in agreement.

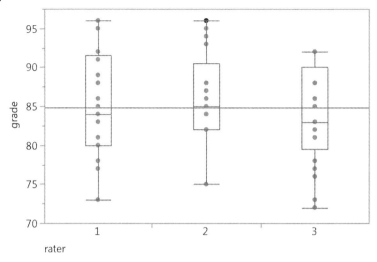

Figure 9.2 Grades by rater.

Create Data Bin and Analyze Spread of the Grades

Part of the data analysis is to develop a regression model (an equation) that predicts the grade based on the four rankings. A good regression analysis needs a wide range of scores rather a tight range of scores. For instance if essentially all of the papers were give a B, it would be difficult for a regression analysis to determine an equation predicting a grade. Instead, we want to see papers with a spread of both high and low scores (Figure 9.3). Figure 9.2 (a box plot) shows the overall distribution, but a histogram will give a more detailed view of how the grades are spread out.

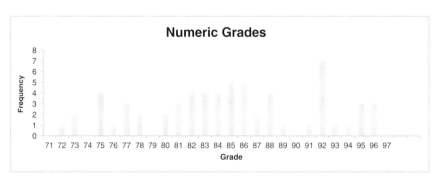

Figure 9.3 Histogram of the grades by numeric scores. There are few papers with any specific score. There is a wide, relatively even spread of grades, which is good for creating a valid regression.

	A	B	C	D	E	F	G	H	I
1	paper	rater	Section	grade	letter	grammar	argument	paragraph	sentence
2	1	1	103	85	B	3	2	4	4
3	1	2	103	88	B+	3	3	5	5
4	1	3	103	82	B-	2	3	3	2
5	2	1	103	78	C+	1	2	3	2
6	2	2	103	75	C	2	2	2	2
7	2	3	103	72	C-	1	1	1	2
8	3	1	103	92	A-	5	5	4	5

Researcher needs to create
this column based on
the grade column

Figure 9.4 Partial table of grades with letter grade column.

The raters assigned a numerical grade to the papers, but for the histogram [defined on page 79], we will want to create a new variable to bin the grades by letter grades (partial spreadsheet shown in Figure 9.4). The result would be Figure 9.5. The scores show a more or less bell-shaped curve centered on B/B−. We also have several papers at each grade, so we have not obvious reason to expect problems with the regression analysis. The skewed curve [defined on page 68] is expected for paper grades, but it does not affect the regression. The regression analysis would be fine even if there were an equal number of grades in each bin. Actually, when doing a regression analysis, it is better to have an equal number data points—in this case, grades—in each of the possible values.

Compare Letter Grades by Section

We started the exploratory analysis looking at box plots of the overall grades by sections. We concluded they appear similar. Now we need to extend that

Figure 9.5 Overall paper grades converted into letter grades. As desired for this study, the grades show a wide spread.

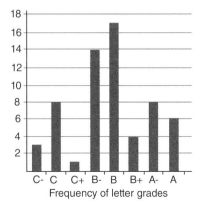

Frequency of letter grades

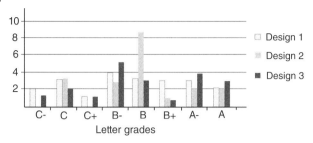

Figure 9.6 Letter grade by section.

analysis to examine if there are differences in the letter grades between the sections. Did one section have more A or C papers than the others?

In this study, we want all three to be similar, although in many studies there may be reasons to expect differences. For example, perhaps each section used a different teaching method and the study is part of the analysis to determine if the different methods result in different grades.

Figure 9.6 shows the distribution broken out by section. Each section received a similar number of grades per letter grade except for the B+ and B grades. There we can see section 103 had an excess of B+ grades and section 142 had many more B grades. Also, notice that section 142 did not have any C+ or C− minus grades, but the other two sections only had a few, so the lack of scores there is not unexpected. There is always random variation between different groups—the basic reason for inferential statistics is to distinguish between random variation and real changes—so having some sets of no values is ok if it is consistent with few values in the other groups.

Compare the Letter Grades by Rater

The last part of the exploratory data analysis is looking at the letter grades by rater (Figure 9.7). We already have looked at the overall scores by rater and now we extend that to look at the individual letter grades they assigned.

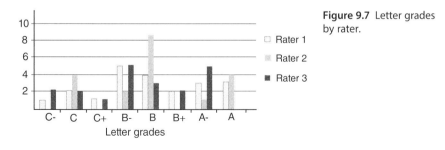

Figure 9.7 Letter grades by rater.

The study used multiple raters to ensure consistency of scores. Thus, we expect that all three raters should give the same number of each letter grade. Unfortunately, it is clear that although raters 1 and 3 are giving similar grades, visually their grades appear to be different from rater 2. In the data analysis, we will perform a statistical test to determine if these differences are significant.

In a more complete analysis, we would also look at the spread on each paper by rater. Because of the calibration procedures, all raters should agree or at least be within a plus/minus grade of each other. With the visual difference for rater 2, that will probably not be the case here. The inter-rater reliability may not be a poor as it seems at first glance; notice that rater 2 seems to have given mostly whole letter grades and few plus/minus grades. Thus, rater 2 might be within the acceptable half grade range of the other two graders. The inter-rater reliability calculations will show how closely they agree.

Calibration of people rating anything

The differences between raters 1 and 3 when compared to rater 2 show the importance of both calibrating the raters before doing a study and using multiple raters.

The calibration routine of having multiple raters practice until they agree helps to minimize the amount of variation that could arise from individual differences. Consider that what people think is poor grammar can vary widely, but is a factor of the individual rater and not of the paper being evaluated.

Multiple raters prevent an individual's view skewing the study results. A group of unbiased people will give scores around the actual mean—have enough raters and you will get a normal curve of their scores.

Unrealistic, but if the study used a large number of raters (such as 1000), the grades given for each paper would form a normal curve.

Inferential Statistical Data Analysis

Inter-Rater Reliability

With three people providing scores, we need to test how closely they agreed [defined on page 87]. Since they all graded the same papers, ideally they should have all given them the same grades and rankings to each paper. However, no group of people will achieve that ideal, so we need to test to see how far from complete agreement they really were. A low inter-rater reliability seriously compromises the results of a data analysis.

This study uses Fleiss's kappa [defined on page 90] for the three graders to see if they were giving consistent grades. If a study uses two raters, then Cohen's Kappa is used, instead.

We find the Fleiss kappa for 3 raters = −0.0129 SE = 0.0314 (calculated from http://www.stattools.net/CohenKappa_Pgm.php).

The SE value is what we use as the test of the agreement. Note that the kappa value is not interpreted the same as a *p*-value on other statistical tests. Conventionally, a kappa of <0.2 is considered poor agreement, 0.21–0.4 fair, 0.41–0.6 moderate, 0.61–0.8 strong, and more than 0.8 near complete agreement. Thus, we can see that there is only fair agreement.

In most studies, fair agreement among the raters is considered a negative. The study report should give the Kappa value and discuss its significance. The pregrading calibration should have resulted in a better kappa.

Cohen's kappa

Cohen's kappa is used to compare the agreement between two people.

Cohen's kappa could also be used if an independent grader was reviewing papers graded by the instructor. Assuming the independent graders have been calibrated to department standards, the Cohen's kappa of the grades they each assign should show a high agreement.

Cohen's kappa (or Fleiss's kappa) are also be used during each round of the grader calibration process to measure agreement. The calibration should continue until the kappa reached a predetermined value.

In the exploratory data analysis, we found the rater 2 consistently gave grades that differed from the other two raters (Figure 9.2). We can check if it was rater 2 that caused the fair agreement ranking by doing a Cohen's kappa on all three pairings of the three graders (1–2, 1–3, and 2–3).

Table 9.1 shows the results for calculations that consider both the plus/minus grades (C−, C, C+, etc.) and single letter grades (C, B, A). A higher agreement for the single letter grades would be expected, since a B and B+ would be considered a disagreement with plus/minus grades, but both become a B with single letter grades.

From the exploratory analysis, we had expected rater 2 to be the problem, but the Cohen's kappa analysis does not bear this out. We should perform a closer

Table 9.1 Cohen kappa results comparing the three paper raters.

	Raters 1–2	Raters 1–3	Raters 2–3
Cohen's kappa for plus/minus letter grades	0.1486	0.0455	0.1407
Cohen's kappa for single letter grades	0.7605	0.5333	0.4432

analysis. What we may find is that although it looks like raters 1 and 3 gave similar scores, they may have given them to different papers. In other words, perhaps they were consistent in papers that rater 1 gave an A− received a B− from rater 3 and vice versa.

We should also look at the inter-rater reliability of the four categories that went into the regression. The analysis for this would be the same as we we have done for comparing the grades. If we look at all five comparisons—the four categories and the paper grade—we may get a better feel for how close the agreement was or what was driving the points of disagreement.

The study started with a grader calibration to prevent low inter-rater reliability. Unfortunately, it did not seem to help in this study. If we had found that a rater was significantly different, as we expected to find rater 2, we would have faced with a couple of different choices.

- Drop different rater and only compare the results of the other two. This is essentially what is done in many sporting events, such as diving and figure skating, where the high and low judge's scores are discarded and remaining scores are used in the final score.
- Continue with all three raters and carefully examine the results to see if there was any skewing.

The second choice of using all three rater is the best one unless there is a substantial and justifiable reason to remove a rater. For example, the rater started to exhibit a negative attitude during the evaluation process (good match for the first half and then shifted to be consistently lower/higher for the second half). Dropping a rater just because it gives better analysis results is not a valid reason and is considered bad research methodology. Whenever a rater is dropped, the reasoning and justification must be carefully laid out in the research study paper. It would be ethically questionable to drop a rater and then write up the results as "two raters gave scores . . . ." without mentioning the third rater.

Grades by Section

We need to test if the grades are different by section. In the exploratory analysis, we saw some differences within the individual letter grades (Figure 9.6), but the overall distribution of scores appears similar.

Because there are three sections, we need to do an ANOVA [defined on page 94] to test if they are different.

First, we will test that the variances[defined on page 39] between the sections are equal since the ANOVA makes that assumption. All four tests calculated by JMP show nonsignificant p-values [defined on page 19], so we can assume equal variances (Figure 9.8).

Level	Count	Std Dev	MeanAbsDif to Mean	MeanAbsDif to Median
103	21	7.255868	6.049887	6.000000
142	21	5.425250	3.968254	3.952381
164	21	6.949135	5.795918	5.619048

Test	F Ratio	DFNum	DFDen	Prob > F
O'Brien[.5]	1.4130	2	60	0.2514
Brown–Forsythe	1.6279	2	60	0.2049
Levene	2.0245	2	60	0.1410
Bartlett	0.9034	2	.	0.4052

Figure 9.8 Test for equal variance across the sections.

The ANOVA gives a p-value equals 0.75, so we can conclude the differences are not significant (Figure 9.9). This is the result we wanted since it means we can combine the three sections for the regression test.

Regression

A regression test gives us an equation that predicts the grade based on the four rankings [defined on page 92].

We have a total of 63 papers in the sample, but we are going to randomly pick two-thirds (42 papers) of them to use to develop the regression equation. The remaining third will be used to verify the equation results.

Running a regression in Excel for the four categories against the grade, we obtain the output of Figure 9.10. The p-value for all four categories is significant, so all four will appear in the final regression equation. If any of terms were not significant, then they would not appear in the final equation. The coefficients for

Source	df	Sum of squares	Mean square	F Ratio	Prob > F
Section	2	24.9841	12.4921	0.2875	0.7512
Error	60	2607.4286	43.4571		
C. Total	62	2632.4127			

Level	Number	Mean	Std error	Lower 95%	Upper 95%
103	21	83.9524	1.4385	81.075	86.830
142	21	85.3333	1.4385	82.456	88.211
164	21	85.2381	1.4385	82.361	88.116

Figure 9.9 ANOVA result showing nonsignificant differences across the sections.

SUMMARY

Regression statistics	
Multiple R	0.947701292
R Square	0.898137739
Adjusted square	0.887125603
Standard error	2.109879278
Observations	42

ANOVA

	df	SS	MS	F	Significance F
Regression	4	1452.27	363.07	81.56	7.83687E–18
Residual	37	164.71	4.45		
Total	41	1616.98			

	Coefficients	Standard error	t Stat	p-Value	Lower 95%
Intercept	66.90	1.27	52.85	0.00	64.33
Grammar	2.62	0.43	6.09	0.00	1.75
Argument	1.00	0.36	2.78	0.01	0.27
Paragraph	1.47	0.46	3.20	0.00	0.54
Sentence	0.86	0.36	2.41	0.02	0.14

Figure 9.10 Regression analysis relating the four categories assigned by the raters against the final grade. The coefficients column gives the values that go into the equation for the *p*-values that are significant. In this case, all values are significant.

the equation are listed under the "Coefficient" heading, which gives a regression equation of

$$\text{grade} = 2.62(\text{grammar}) + 1.00(\text{argument}) + 1.47(\text{paragraph})$$

$$+0.86(\text{sentence}) + 66.90$$

Based on this equation, if we know the four values, we can estimate what a student's grade will be on a paper.

When all terms are not significant

The example in Figure 9.10 had all four terms showing significance and thus they all appeared in the regression equation. If the results had looked like Figure 9.11, there are two terms that do not show significance—the shaded cells. Because they are not significant, the do not contribute to the grade.

	Coefficients	Standard error	t Stat	p-value	Lower 95%
Intercept	66.90	1.27	52.85	0.00	64.33
Grammar	2.62	0.43	6.09	0.43	1.75
Argument	1.00	0.36	2.78	0.01	0.27
Paragraph	1.47	0.46	3.20	0.23	0.54
Sentence	0.86	0.36	2.41	0.02	0.14

Figure 9.11 Regression analysis relating the four categories assigned by the raters against the final grade. Only two of the categories have significant values.

Running a regression multiple with different random sets

We created the regression with a random choice of 42 papers (out of 63 total). If we rerun the regression with a different set of 42 papers, the equation coefficients will be slightly different. As a check we can run it multiple times with a different randomly chosen set of 42 papers and compare that they are all similar. The table shows the coefficient and p-value results of four different runs with 42 randomly chosen papers. In reality, if we run enough times, the values will produce a normal curve with a mean approximately equal to the real mean.

The coefficients generally agree with across all five runs—including the first one listed in Table 9.2—but notice that the sentence value is only significant some of the time. Further analysis would be required to understand what is happening here.

Table 9.2 Repeating regression with subsets randomly drawn from the study data.

	Run 1		Run 2		Run 3		Run 4	
	Coef.	p-Value	Coef.	p-Value	Coef.	p-Value	Coef.	p-Value
Intercept	67.44		67.38		66.82		66.54	
Grammar	2.90	<0.001	2.78	<0.001	2.69	<0.001	2.77	<0.001
Argument	1.36	0.003	1.23	0.004	1.30	0.001	1.10	0.015
Paragraph	1.24	0.014	1.57	0.001	1.43	0.003	1.33	0.003
Sentence	0.31	0.393	0.30	0.353	0.50	0.192	0.78	0.022

Figure 9.12 Residuals plot for the grade equation. The even number of points above and below the line is what is desired. However, the overall scatter is higher than desired; this means the equation will probably not give a reliable prediction.

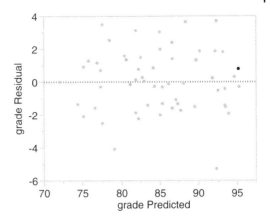

Residuals Plot

Residuals [defined on page 210] show how closely the points fit the equation.

A simple test for how well a regression equation fits the data can be determined by a residuals plot. This plot looks at each of the points in the data set and determines how closely they fit the equation (Figure 9.12). The residuals should be randomly placed both above and below the regression line and should also be close to the line.

The graphical fit of the line can tell you if the regression equation is valid. It can also tell you if a different type of equation is needed. This study used a linear regression (a straight line); if the plot showed a close fit over part of the line and then an increasingly poor fit, the actual data may need a quadratic or exponential regression. If the data needs this type of regression, consult a statistician.

Verification of a Regression Equation

The regression equation will be valid for the papers that were used for the regression analysis; however, it does not necessarily apply any others. Since the goal of the study is to produce a more general equation, it is important for a study to verify the equation as part of the final report.

When we created the regression, we randomly picked 2/3 of the values to create the regression equation. We do this because we want to be able to verify the regression. If we had used all of the values, then any data point we used would fit the regression equation; in any model, points used to create the model cannot be used to verify it.

Verifying the regression (or any model) is important. When you run a regression, it will give a result since there is a best-fit line through any set of points. However, the result may be more/less meaningless. Part of a research project is to ensure the results actually have meaning; thus, the researcher must hold back some of the data points and see if they also fit the model.

In this study, 42 papers were used in the analysis. The remaining 21 papers, not part of the 42 used to determine the regression equation, will be used to test its validity.

The 42 papers that were used to create the regression equation cannot be used to evaluate it. There should always be a high correlation between the estimated and actual score for those 42 papers, but that does not ensure the correlation will carry over to other papers (or in general terms, other points in the data set).

The verification papers' values are plugged into the regression equation and the resulting grade estimate compared against the actual grade (Figure 9.13). Using JUP Fit Line (or any other test that compares goodness of fit) can be used to test the correlation between the actual grade and the regression-estimated grade.

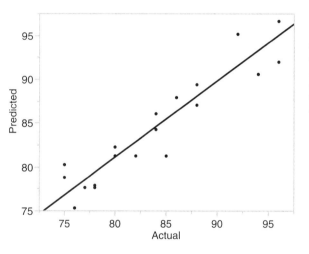

Figure 9.13 Test of the regression equation. The data points should all plot on the line. The results show the regression equation fits poorly. Note the Prob > F term—the p-value— is <0.001.

Analysis of Variance

Source	DF	Sum of Squares	Mean Square	F Ratio
Model	1	765.59306	765.593	154.3117
Error	19	94.26547	4.961	**Prob > F**
C. Total	20	859.85852		<.0001*

Parameter Estimates

Term	Estimate	Std Error	t Ratio	Prob>|t|
Intercept	11.601173	5.938102	1.95	0.0656
Actual	0.8692879	0.069978	12.42	<.0001*

Unfortunately, the low *p*-value (<0.0001) shows there is not a good fit for the verification papers to the regression line. A perfect fit would have all of the points on the line. The more the scatter—higher above or below the line for each point—the weaker the fit. It seems the data collected in this study would not be a good predictor of the student's final grade.

The regression in this study used for numeric grade. If the regression was rerun with the letter grades, the fit might be better. In that case, it would find it could predict a B+ grade, but not distinguish between the 87 and 89 score.

Deeper analysis than the initial equation

Human performance in selection tasks is frequently described by Fitts' law, which states that the average time needed to move to a target and select it is related to the distant to the target and its width. Fitts' law is relatively old and has been a staple of interaction design for over 50 years.

Recent work (Nieuwenhuizen and Martens, 2016) examined some of the factors and proposed two generalizations of Fitts' law. The actual work done by Nieuwenhuizen and Martens is not relevant to this book, but the idea of the deeper analysis is. If a study finds a basic relationship, such as Fitts' law, the researcher has two choices: (1) call that relationship good and stop or (2) do a deeper analysis. Granted the deeper analysis probably is a new study, with more participants and a design focused on the specific factors that the researcher hypothesized are relevant.

Research is a trajectory, not a series of on–off studies.

Exercises

Student Questions

1) A study used four people to evaluate the difficulty people had performing a task. A Fleiss' kappa showed a low inter-rater reliability (value 0.15). Explain how to handle this as part of the study analysis.

2) Define residuals and explain why they must be considered as part of interpreting a regression?

3) If we had run a regression for each rater (in other words, ended up with three equations . . . which would be similar to only having one rater in the study), would we expect the regression equation to give a better or worse prediction than using a combined value of all three raters? Why?

Table 9.3 Regression output results.

SUMMARY OUTPUT

Regression Statistics	
Multiple R	0.856955
R Square	0.734371
Adjusted R Square	0.747126
Standard Error	3.209879
Observations	42

ANOVA

	df	SS	MS	F	Significance F
Regression	4	1452.27	363.07	81.56	3.69E-13
Residual	37	164.71	4.45		
Total	41	1616.98			

	Coefficients	Standard error	t Stat	P-value	Lower 95%	Upper 95%
Intercept	17.80	1.27	52.85	0.00	64.33	69.46
Speed	8.35	0.43	6.09	0.64	6.25	9.64
Practices	2.10	0.36	2.78	0.02	1.10	3.58
Pretest	5.35	0.46	3.20	0.00	3.78	8.99
Extra work	6.24	0.36	2.41	0.08	4.36	8.04

4) If the results of a regress in Excel are as provided in Table 9.3, what is the regression equation?

Data Analysis

1

Perform the exploratory data analysis and the inferential statistical analysis for one of four categories (for example, argument). You will have to graph the data spread, calculate the Fleiss kappa, and evaluate if there are statistically significance differences between sections.

2

Perform the regression using the letter grades. (You will have to convert them to numbers.) How does the result of the regression change from the numerical grade? Do the residuals increase or decrease?

Additional Data Sets

Class factors affecting final grade. Data in: Chapter 9 class-teaching-study.xlsx

Reference

Nieuwenhuizen, K. and Martens, J. (2016) Advanced modeling of selection and steering data. *International Journal of Human-Computer Studies*, **94**, 35–52.

10

Specific Analysis Examples

The chapter explains the issues that must be considered for specific problems or issues that need to be considered in many data analysis. Rather than a detailed analysis, this chapter focuses on a single data analysis issue, discusses it conceptually, and explains how to handle it within the data analysis.

Handling Outliers in the Data

Outliers are values that do not appear to fit within a data set because they are too low or too high compared to the rest of the data.

General data analysis guidance is to not ignore or discard outliers unless there is a clear and justifiable reason to do so. The data analysis can also perform the analysis both before and after handling the outliers and then the results can then be compared. Also, the study report should discuss how and why outliers were handled within the data analysis—whether they were removed, adjusted, or used as is.

If a data set has multiple outliers or it seems have a wide spread, it is best to consult a statistician. Multiple outliers and wide spreads tend to indicate the data do not have a normal distribution, but instead has very heavy tails. The issue can be resolved by any of several formal statistical tests for outliers, but they are best done by a statistician who understands those tests' assumptions and limitations. In addition, depending on the data spread, specialized non-parametric tests may be required.

Two populations can appear as outliers

A study that inadvertently uses at two different populations without clearly distinguishing them can easily have data that appears to contain outliers.

Introduction to Quantitative Data Analysis in the Behavioral and Social Sciences,
First Edition. Michael J. Albers.
© 2017 John Wiley & Sons, Inc. Published 2017 by John Wiley & Sons, Inc.
Companion website: www.wiley.com/go/albers/quantitativedataanalysis

Consider if a study randomly selected people at a gym to look at maximum bench press strength and picked 10 men and 2 women. The women's maximum bench press weight will almost certainly look like outliers, but they are actually part of a different distribution.

With a large enough sample, the graph would appear bimodal, suggesting a mix of two populations. But the smaller sample sizes in the social sciences often do not have enough points to show this clearly.

Various factors that can cause outliers are the following:

- **Chance Event in the Data Distribution:** By chance, the study found a four-standard-deviation event. This should remain in the data for analysis.
- **The Study Sample Is Bimodal:** If most of the subjects are from one group, some of the other group may appear as outliers. (See the sidebar: Two populations can appear as outliers.)
- **The Population Has Long and Heavy Tails:** The population does not have a normal distribution, and events that seem well away from the mean/ medium are common. Populations with heavy tails or that are skewed can give lots of data points that appear to be outliers, but are actually within the expected frequency distribution. Since these distributions are not a normal distribution, use of statistical tests that assume normality must be reconsidered with analyzing the data.
- **Noisy Data:** The data itself might have a high variance, and even repeated trials with the same subject can give widely different results. Noisy data tends to have a high standard deviation so, even if should fit a normal distribution, a small data sample may appear to have extreme outliers that fit within two standard deviations.
- **External Factors That Are Not Controlled for in the Study:** For example, one subject was very tired or not as computer-savvy as they claimed.
- **Data Measurement Error:** A value was recorded wrong. Clearly, this should be removed before the data analysis. Unfortunately, it is often difficult to know if the point is a true outlier or was a measurement error.

Effects of outliers

Calculation of mean values is very sensitive to outliers. If a study of reading speeds found values of 54, 134, 143, 153, 155, and 163. The value of 54 is an outlier. The subject may be a very fast reader, did not really read the material, or the timing method may have been flawed (perhaps the stop watch was stopped early). The problem is the average reading speed is very different with and without the outlier: 54, 134, 143, 153, 155, 163

With: $(54 + 134 + 143 + 153 + 155 + 163)/6 = 133.67$. Note the average reading speed is less than the speed of all by one subject.

Without: $(134 + 143 + 153 + 155 + 163)/5 = 149.6$

Whether or not to reject the 54 as an outlier requires the researcher to reexamine the data and determine if it makes sense for that value to belong within the data set. Should that fast reader have been in the sample? Was there a data collection error—a 154 value was recorded as 54? You could examine that person's reading comprehension test score; if it that was approximately chance, then that person clearly was not reading and the point can be safely dropped.

Handling Outliers in Data

When the data has outliers, a researcher has to decide how to handle them at the beginning the data analysis. It is important to make the choices about how they will be handled early. Running and rerunning statistical tests using different methods of coping with outliers until significance is found compromises the integrity of the analysis. Manipulating the data changes the confidence intervals. Data that has significant skew will be strongly affected by any manipulation since it will tend to reduce the overall skewness.Leys et al. (2013) looked at how different research articles handled outliers and are critical of the methods used in those studies. They point out that many of the methods do not seem to reflect an understanding of handling them and the details of the method used are not reported in the article.

The first step in cleaning up the outliers in the data is to examine the individual points and try to determine what caused the value. Each outlier needs be examined individually and its inclusion/deletion justified. A valid data point must be retained and a bad data point (perhaps a timer problem or the subject not understanding the directions) can be discarded. There is also the chance that the data contains two different distributions and the outliers fit well within one of those distributions.

Methods for detecting outliers in the data can include

Standard deviation multiplier — If the data fits a normal distribution, points more than a value of the SD, typically 2.5 SD or 3 SD are considered outliers and are dropped.

In a normal distribution, three standard deviations is 99.7%, so data with 300+ samples has a reasonable probability of having a data point greater than 3 SD from the mean. For data with a large standard deviation, data points that look

Box plots

like outliers may be within 3 SD of the mean. A problem with using the standard deviation method is that its value is sensitive to the outliers or that the data may not fit a normal distribution.

Data that do not fit a normal distribution should be graphed with a box plot. Different software programs produce different box plots that show the tails and the outliers differently. But they all do give a visual method of deciding where to set the cutoff for declaring a point an outlier.

Methods of Handling Outliers

Do Nothing

Analyze the data with the outliers. Nonparametric statistics is often the best choice since by their nature they negate the outlier's effect. Parametric statics are sensitive to outliers, but the nonparametric tests might limit the overall data analysis.

It is important to notice that during the random sampling, there were some means that fell outside of the 5 and 95% range. This is normal and expected. In a small data set, a researcher might be tempted to toss them as outliers. However, there is a chance that the experimental data returned a value outside the two standard deviation range.

There is also a chance that the data set itself is not normally distributed. If the data has a flatter curve than a normal distribution, what could appear to be outliers are actually points well within its expected values. You can test for normal distribution, but with small data sets, these tests are not very robust. Also, the outliers may cause the normality test to reject the data as is normally distributed.

Delete the Outliers

Analyze the data after removing the outliers. Removing data is rarely a good choice since it introduces a selection bias. Data points should only be removed if there is a strong justification and that justification must appear in the study report.

One method of deleting outliers is to use fixed cutoffs, such as 5 and 95% percentiles (2 SD value) or 0.03 and 99.7% (3 SD value); all data outside of these are removed. Doing so have removed all data outside of two or three standard deviations. This method does have the advantage that all low/high values are removed and the study cannot be accused of cherry-picking data point removal.

Figure 10.1 Box plot.

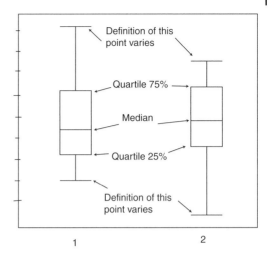

A graphical method is to use box plots with fences (NIST/SEMATECH) to set the cutoff points. All data outside the fences must be examined. This explanation assumes you are adding the fences to a standard box plot. Depending on the software, it may calculate them for you.

After creating the basic box plot (Figure 10.1), calculate the interquartile range (IQ) as the difference (Q3 − Q1).

The fences are the drawn in as follows:

- lower inner fence: $Q1 - 1.5 \times IQ$
- upper inner fence: $Q3 + 1.5 \times IQ$
- lower outer fence: $Q1 - 3 \times IQ$
- upper outer fence: $Q3 + 3 \times IQ$

A point between the inner and outer fence is considered a mild outlier and a point beyond an outer fence is considered an extreme outlier. The data analysis can be done with deleting just the extreme outliers and then both sets of outliers.

Bin the Data

The data can be restructured from interval into ordinal data by creating a set of bins, such as very low, low, average, high, and very high. Moves the outliers into the very low and very high bins. Once this data is restructured, use non-parametric statistics for the analysis.

Clearly, creating the data bins changes the data significantly by converting it to ordinal data. This changes the effect size calculations and the power of the tests.

Transform the Data

Transform ratio data with a mathematical transformation (nominal and ordinal data should not be transformed) (Bland and Altman, 1996). Each data point is converted to a transformed valued: $x_i = f(z_i)$. The advantage of a good transformation is that it tends to move the entire data set closer to a normal distribution. If the data requires transformation, consult a statistician.

> ### Outliers are different from floor/ceiling effects
>
> Outliers are data points that do not seem to fit into the overall data set. Floor/ceiling effects are when a substantial part of the data is clustered at the bottom/top, such as all test scores of 95–100. Poor data analysis sometimes confuses them.
>
> Outliers will appear in valid data collection and need to be handled. Floor/ceiling effects are an indication of data collection methodology and compromise the fundamental data quality.

Floor/Ceiling Effects

Floor and ceiling effects are when the data clusters at either the top or bottom of the allowable range. They can be considered essentially the same, but at opposite ends of the data set.

- A floor effect is when most of the subjects score near the bottom.
- A ceiling effect is when most of the subjects score near the top.

A floor/ceiling effect occurs when a measure possesses a distinct limit for potential responses and a large concentration of participants score at or near this limit. The research problem is that the effect of the any changes to the dependent variables becomes more/less impossible to detect for relatively large changes to the independent variable.

Examples include the following:

- The questions are too hard for the group you are testing. This is even more of a problem with multiple-choice tests, where because of guessing, the scores will be about 25%. Or the test is too easy and a large percentage of the subjects answered all of the questions correctly. Now you cannot tell if the independent variable has changed the dependent variable.
- Survey questions result in most people picking the highest or lowest value. Customer satisfaction surveys tend to suffer from this problem.
- Task time tests where most people complete the task in the minimum time. This might also be a data collection issue, such as timing the speed to click on

three targets with a mouse. If the recording device is not measuring in milliseconds, everyone's time might look the same.

A floor/ceiling effect causes your data to become binary—or can easily be collapsed into two categories without much loss of information—either within the effect group or not in the effect group.

Floor or ceiling effects and performance asymptotes

Floor and ceiling effects are related to, but different from, performance asymptotes. Asymptotes occur when subject's score cannot exceed a specific value. Response time studies have found people need 250 milliseconds to press a key. Or, in pre/post tests, a score cannot be more than 100%. Especially in timed studies, this could represent a natural physiological limit—a performance asymptotes —rather than an artifact of the measuring devices or the study. In this case, the limit becomes an interesting finding of the study.

Consider how, for a timed task, the researcher who knows how to complete the task takes 15 s. This can be considered the fastest possible time—a lower performance asymptote—for completing the task. A cluster of times ranging from 16 to 18 s could indicate a floor effect in the study.

Order Effects

Order effects are when a study can get different results based on the order in which the subjects perform the control and experimental conditions. For example:

- Later exam questions might be answered more quickly/correctly because they were primed by earlier questions. (Think of the times you remembered/changed an earlier exam answer after reading a later question. That is an order effect.)
- Subjects get bored with a long list of questions or tasks and fail to give them full attention.
- Tasks may be completed quicker because of the practice gained by earlier tasks.

Consider how the order effect changed the results of this data analysis problem.

A set of exam problems might be completed more quickly in one order than another, because one problem might prepare you for another but not vice versa. So if a control group of students is presented test problems in a "good" order and the treatment group gets the problems in a "bad" order, then a positive effect of treatment might be washed out by the effect of order; or, if the treatment group gets the "good" order, the effect of treatment (which could be zero) might appear larger than it really is. (Cohen, 1995)

The study design has to allow for order effects from the early design, with techniques such as using independent measures or counterbalancing. Once they are introduced into the data, they are almost impossible to allow for as part of the data analysis.

Data from Stratified Sampling

A stratified sample is a sampling technique in which the target population is divided into different subgroups and then the subjects are randomly selected proportionally from the different groups. The purpose is to ensure there are enough people in each group to provide enough statistical power for valid results. Texts on sampling methodologies should detailed descriptions of how to set up a stratified sample. One issue is that it can create lots of cells in the table and require the availability of a large sample.

With stratified sampling, the data analysis should look at both the larger groups and the individual subgroups.

Assume a study looked at students and grouped then according to the amount of exercise they did. Other data such as class, type of exercise, and physical condition were also collected (Table 10.1). The analysis goal is to look at and compare each of the subgroups against each other as determined by the study's hypothesis. For example, as a first pass at the analysis, test by class (freshman, etc.) with all exercise levels combined. Then, test each subgroup, such as freshman-moderate exercise versus sophomore-moderate exercise and freshman-moderate exercise versus freshman-high exercise.

Missing Data

Missing answers are part of life in social science research. For example, some people do not answer questions on a survey or, in observational studies, some selected observations may not be recorded either because they did not perform an action or researcher error in recording it.

There are many ways that survey data can be missing and each one needs to be handled differently (Acock, 2005). Handling the missing data is not really a data analysis problem so much as a methodology problem. Before the data are collected, you need to define how missing data will be handled. Sometimes the missing value can be ignored and the rest of the data for the person is ok and other times missing values might negate part or all of the data for that subject.

Table 10.1 Stratified sample for different types of exercise.

			Subjects needed
Freshman	High physical condition	Running	A close to equal number of subjects in each of these cells.
		Bicycling	
		Swimming	For good test power, we may need 10–20 people (or more) in each cell. A total of 180–200 subjects.
	Average physical condition	Running	
		Bicycling	
		Swimming	
	Poor physical condition	Running	
		Bicycling	
		Swimming	
Sophomore	High physical condition	Running	
		Bicycling	
		Swimming	
	Average physical condition	Running	
		Bicycling	
		Swimming	
	Poor physical condition	Running	
		Bicycling	
		Swimming	

There has to be enough people in each of the right-hand cells (with exercise type) to support the statistical analysis. Ideally, a close to equal number in each cell. This shows how quickly the number of total subjects required can grow with multiple subconditions.

There are many ways that data can be missing and each one needs to be handled differently. A few common ones are the following:

- If the survey is administered at two different times, some people may only complete the first one. They took the first survey, but did not retake it 2 weeks later.
- People will not answer some questions, either intentionally or unintentionally. Some online surveys eliminate this issue by forcing answers to all questions, but that creates its own methodological problems.
- Values may be missing because the person was told to skip them. For example, when a question says "if you answered NO, skip to question 14." Then several questions have no answer. It is intentional in the design, but it must be handled in the analysis.

If you need to use any of the methods to replace missing data, you need to enlist the aid of a statistician or someone skilled in handling missing data. Deriving the correct algorithm to calculate the missing values is nontrivial. Since you are inserting data, there is a potential for confounding the results.

Replacing missing data with a weighted response

There are some missing data methods that work to replace the missing values by weighting the responses and calculating the likely value (Brick and Kalton, 1996).

For example, survey question 5 is blank and 80% of the respondents who answered questions 1–4 the same as the person who did not answer 5, answered question 5 with D. Thus we assume that person would have answered question 5 as D. Therefore, we will replace the missing answer with D. Note that this method is *not* a simple 80% of the question 5 answers are D, so replace the missing answer with D. Instead, it looks at how people with similar responses answered the question.

Any method of replacing missing data must be fully described in the study report.

Noisy Data

Noisy differs from outliers because it affects every data point. It can be considered as a random value added to every data point.

Social science data that involve people engaged in real-world activities—as opposed to a laboratory setting—is almost always very noisy. The people are engaged in other things other than what you are recording and that influences how they are acting or responding to the situation.

Causes

- Poor quality equipment
- Uncontrolled or unaccounted for factors in the study
- Inherent in the process being studied. The number of text messages sent per hour will naturally has a wide spread.

How to handle

- Transform the data. This might involve some complex functions and will require the assistance of a statistician.
- Redo the study with a new methodology that considers the previously uncontrolled factors. A partial data analysis should uncover the uncontrolled factors; a purpose of a pilot study is to help uncover these factors.
- Perform the analysis with nonparametric statistics using the assumption that the noisy is inherent in the data.

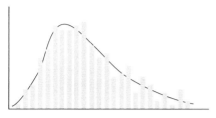

Figure 10.2 Graph after a log transformation. The skewness of data has been removed and the data appear to follow a normal distribution.

Transform the Data

Data that is noisy, skewed, or has a large standard deviation may benefit from a data transformation (nominal and ordinal data should not be transformed) (Bland and Altman, 1996). With a transformation a new value is calculated for each data point, $x_i = f(z_i)$. Unlike creating data bins, parametric statistical analysis can still be used.

The advantage of a transformation is that it tends to move the entire data set closer to a normal distribution (Figure 10.2). However, the transformation does change the data and overall test power is typically decreased. Data analysis that needs a transformation should use it, but it should not be indiscriminately used in a search for finding statistical significance in the data.

Performing different transformations and rerunning statistical tests until one shows significance is poor data analysis. If the data requires mathematical transformation, consult a statistician.

Another potential transformation method is normalizing the data. In a business analysis, this might mean converting stores total sales to sales per square foot so that different sized stores can be directly compared.

Any transformation must make sense in its final form. Sales per square foot is a logical unit for business analysis, but a transformation that results in task time per reading level does not.

If a data transformation is used, the study report must explain and justify the transformation method.

Mathematical transformation of the data

A common transformation is taking the square root of all of the values (or taking the natural log of all values). What this means is that instead of running a statistical test on the data values, the test would be run using the transformed values.

Data value		Square root transformation		New value
4882.2	→	sqrt(4882.2)	→	69.9
9548.0	→	sqrt(9548.0)	→	97.7
4895.5	→	sqrt(4895.5)	→	70.0
4761.4	→	sqrt(4761.4)	→	69.0
5562.8	→	sqrt(5562.8)	→	74.6
7287.2	→	sqrt(7287.2)	→	85.4

The six data values in the left are the date collected in the study. A square root transformation is applied that yields the value in the right-hand column. The statistical test would be run on these new values.

References

Acock, A. (2005) Working with missing values. *Journal of Marriage and Family*, **67** (4), 1012–1028.

Bland, J.M. and Altman, D.G. (1996) Statistics notes: transformations, means, and confidence intervals. *British Medical Journal*, **312**, 1079.

Brick, J.M., and Kalton, G. (1996) Handling missing data in survey research. *Statistical Methods in Medical Research*, **5**, 215–238.

Cohen, P. (1995). Empirical Methods for Artificial Intelligence. Retrieved from http://www.cs.colostate.edu/~howe/EMAI/ch3/node11.html.

Leys, C., Ley, C., Klein, O., Bernard, P., and Licata, L. (2013) Detecting outliers: do not use standard deviation around the mean, use absolute deviation around the median. *Journal of Experimental Social Psychology*, **49**, 764–766.

NIST/SEMATECH e-Handbook of Statistical Methods, http://www.itl.nist.gov/div898/handbook/prc/section1/prc16.htm (accessed June 13, 2016).

11

Other Types of Data Analysis

There are other types of data analysis that are often used. This chapter briefly introduces them, but does not attempt to explain how to perform the analysis.

Time-Series Experiment

Some of these studies are looking at cyclic data that display some sort of pattern that repeats at over time intervals. This type of data is called a time series.

A time series is a collection of observations of well-defined data items obtained through repeated measurements over time. A times series allows you to identify change within a population over time. A time series can also show the impact of cyclical, seasonal, and irregular events on the data item being measured.

- Number of insects in a plot varies in a cyclic manner by day and season.
- Weight of people's clothes worn varies in a cyclic manner by season.
- People's response time varies in a cyclic manner by time of day.

In the basic form, time-series data resemble a sine curve (Figure 11.1). Part of the study methodology must estimate the curve period and ensure the data collection captures the cycle.

Besides the sine wave element, time-series data may also have a trend line, which is the overall average change in the values (Figure 11.2). This may be increasing, decreasing, or may itself be a longer term sin wave. For example, daily temperature changes over a 24-h period and daily average temperature changes over a year period.

Evaluating trend data requires collecting data over a significant time period to capture the overall trend and ensure it is distinct from the short-term changes.

Time-series data often shows jumps, abrupt, and permanent changes. An experiment on time-series data often causes a jump in the data by changing something either in the cyclic component or in the long-term trend. Measuring

Introduction to Quantitative Data Analysis in the Behavioral and Social Sciences,
First Edition. Michael J. Albers.
© 2017 John Wiley & Sons, Inc. Published 2017 by John Wiley & Sons, Inc.
Companion website: www.wiley.com/go/albers/quantitativedataanalysis

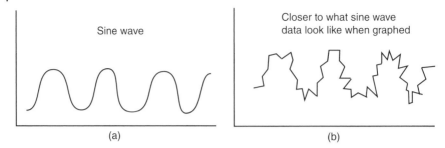

Figure 11.1 Basic time-series data. (a) The graph shows a simple sine wave. (b) The graph shows more realistic data.

Figure 11.2 Time-series data with an increasing trend. Depending on the study goals, either the long-term trend or the short-term variation might be more important.

the extent of those changes is typically the experimental goal. At other times, outside influences may have caused the change. In that case, the analysis needs to figure out the reason and decide if the overall study data are still valid.

The changes in Figure 11.3 could be the following:

- People in a park and then a big snowstorm happened.
- Measure of consumer confidence and then the stock market took a major drop.
- Task time for an activity and then there was a training session on how to do it faster.

Figure 11.3 Time-series data with jump. The experimental intervention or an outside factor has influenced the data and caused an abrupt and permanent change in the data values. Here both the individual data points and the long-term trend slope have changed.

Evaluating time-series data gets complex quickly. It is important consult a statistician during the study design to help determine the length of time data must be collected and during analysis to help analyze the data.

Time Series with Multiple Series of Data

The time-series data collected in a study may contain multiple factors that each vary on their own cycle. There may be daily, weekly, monthly, and yearly cycles all contributing to the data values collected in the study.

The individual sin waves can be parsed out using Fourier analysis (Figure 11.4), which takes a complex wave form and determines the sin waves that make it up. On the other hand, a Fourier analysis always produces a set of sine waves. If the time series contains waves that repeat at varying intervals (do not conform to a sine wave), it may not be useful.

Performing a Fourier analysis in nontrivial; consult a statistician or a mathematician.

Figure 11.4 Fourier analysis. The three sin waves combine to give the top complex wave. Often time-series data consists of multiple cyclic waves like these three.

Collected data graph looks like this

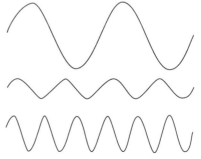

Fourier analysis resolves the complex graph into a series of sine waves

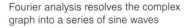

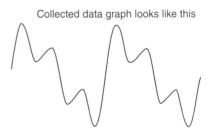

Analysis for Data Clusters

Larger data sets often contain different groupings of data; essentially the data set contains multiple populations. Cluster analysis works to bring out those differences. The individual clusters (populations) can then be examined.

Cluster analysis groups a set of objects in such a way that objects in the same group (called a cluster) are more similar (as defined by the researcher) to each other than to those in other groups (clusters). Cluster analysis divides the data into two or more groups based combinations of variables to determine which properties the data points share. As such, cluster analysis can result in different groupings based on the combination of variables that are chosen. For example, clustering the same group of people by athletic ability (throwing, running, strength, endurance) would result in a different cluster diagram than if they were clustered by other measurements (age, work type, body build, salary, hobbies, etc.).

Figure 11.5 shows one potential cluster diagram for various fruits. A different cluster diagram would result if they were clustered by different criteria (e.g., plant size, area grown in, length of growing season, and type of soil needed).

Most dedicated statistics software programs support cluster analysis and statisticians have developed different methods of clustering data. Consult a statistician for the best method.

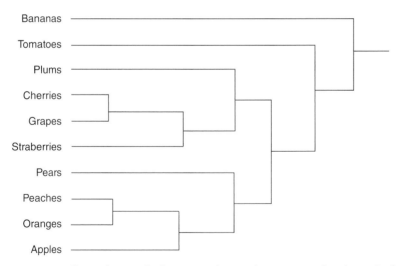

Figure 11.5 Cluster diagram for fruit. Depending on the criteria used to cluster the fruit, the cluster diagram will change.

Low-Probability Events

Some events that need to be studies occur with a low probability, which makes them difficult to study. On the other hand, they may have high consequences. The field of risk management typically works with this type of event: a disaster has a low probability of occurring, but has a very high cost when it does.

Consider a study that looked at why good students did not do well in a specific course, such as a study that looks at why students who performed well in Comp 101 fail to perform well in Comp 102. Clearly this is a small percentage of the students and the reasons for failure are widespread. Most classrooms will not have a student who fits the subject criteria and, even among the applicable students, there will be no single factor.

A statistician is needed early in the study design to help determine how many subjects are needed.

Metadata Analysis

Metadata analysis is a method of taking the data from several studies and reanalyzing the combined data set. They are common in health care where the results of several drug or health intervention studies are combined and analyzed. If there are 27 studies about cardiac surgery education that showed widely different success factors, a metadata analysis that combined all 27 might be able to identify what is happening.

A formal definition:

> A systematic method that uses statistical techniques for combining results from different studies to obtain a quantitative estimate of the overall effect of a particular intervention or variable on a defined outcome; MA [meta-data analysis] produces a stronger conclusion than can be provided by any individual study. (Free Dictionary)

The general aim of a meta-analysis is to more powerfully estimate the true effect size. In other words, was the effect being studied really a dependent variable? If multiple studies reveal that it is, then a meta-analysis can help to clearly establish the effect size.

Advantages
- Greater statistical power
- Precision and accuracy improves with more data
- More confidence in the results being applicable to the general population

Disadvantages
- Difficult and time consuming to identify studies, since the studies must have similar data collection methodologies. In the social sciences, it can be difficult to find a significant number of studies.
- Poor choice of studies can result in useless results. Good analysis of bad data yields a bad result.
- Requires advanced statistical techniques because the variables need to be standardized.

Standardized Variables

There are times when a researcher wants to compare two variables, but they are measured on different scales that make a simple comparison meaningless. In these situations, the values need to be converted to standardized variables (also called the z score). When a study design requires the use of standardized variables, it is best to consult with a statistician to ensure the conversions are done properly.

Standardized variables have a mean equal to 0 and a standard deviation equal to 1. To calculate a standard value:

$$z = \frac{x - \mu}{\sigma}$$

where

μ is the mean of the population.
σ is the standard deviation of the population.

The quantity z represents the distance between the raw score and the population mean in units of the standard deviation. z is negative when the raw score is below the mean, and is positive when above the mean.

Conversion to standard variables is a substantial part of performing a meta-analysis, since values obtained in different studies are being compared. In a meta-analysis, the results of different studies are combined and since each study used different research methods, different scales, and obtained different data spreads, the values often need to be standardized.

Using standardized variables

A study wanted to compare student's freshman entrance tests in writing and mathematics against their writing grades in freshman composition. Clearly, these are all measured on different scales. By creating standardized variables for the sets, the distributions can be compared to determine if they show strong similarities. After being standardized, the data distribution for all three variables will have the same mean (zero) and a standard deviation of 1. The values for individual students can then be used in a regression or other statistical analysis with the assumption that each value is weighted to contribute equally to the overall effect.

Reference

Free Dictionary Definition of meta-analysis. http://medical-dictionary.thefreedictionary.com/meta-analysis (accessed May 12, 2016).

Appendix A

Research Terminology

This section is only giving a barebones description of research method terminology to ensure the reader is using these terms in the same way this book uses them. In general, it assumes the reader is already familiar with them since they are required to design a coherent research study and collect reliable and valid data.

Independent, Dependent, and Controlled Variables

A quantitative study will have a minimum of two variable types: independent variables and dependent variables. The purpose of the study is to uncover the relationships between those two variables.

Independent variable   The variable which is manipulated by the researcher. By controlling the independent variable, the researcher causes the dependent variable to change.

   In the hard sciences and engineering, the independent variable can be manipulated. The electrical voltage or the pressure applied to the system or the wind speed can be precisely changed and the resulting change to some element of the system (dependent variable) can be measured.

   In many social science studies, the independent variable cannot actually be manipulated. For example, if the study wants to see if average hours of sleep per night (independent variable) affect overall college GPA (dependent variable). The study could assign groups and then force people to always sleep a fixed number of hours per night, but human nature and research ethics both prevent that sort

Introduction to Quantitative Data Analysis in the Behavioral and Social Sciences,
First Edition. Michael J. Albers.
© 2017 John Wiley & Sons, Inc. Published 2017 by John Wiley & Sons, Inc.
Companion website: www.wiley.com/go/albers/quantitativedataanalysis

of manipulation of the variable. Instead, the researcher can only collect data on people's sleeping habits. In the end, the analysis groups people with similar sleeping habits and then analyzes them as if they had been assigned to groups with fixed hours of sleep per night.

Dependent variable   The variable which—the researcher hypothesizes—changes as a result of changes in the independent variable. In other words, the dependent variable is the action or event being studied and is expected to change whenever the independent variable is altered.

For example, based on the independent variable (such as hours of sleep per night) the study subjects are divided into multiple groups that are analyzed with respect to the dependent variable (such as overall GPA, reading speed, ability to comprehend healthcare literature, and muscle strength)

Controlled variables   The variables that the researcher attempts to keep constant to prevent their influence on the effect of the independent variable on the dependent variable.

In a social science study where variables often cannot be held constant, these variables may need to be examined as part of the study to determine if they also vary with the dependent variable. The study looking at hours of sleep per night and overall GPA would have to control for factors such as alcohol use, work schedule, social interactions, and college major.

The extent of manipulation the research can do depends upon the study situation. In a laboratory study, the independent variable can be changed and the resulting changes in the dependent variable recorded. In a typical social science study where most variables cannot be adjusted, their values can only be recorded across a range of subjects/situation and then analyzed.

Examples

Compare these two studies and how they can/cannot adjust the independent variable:

- To study if various light levels (independent variable) affect people's reading rate (dependent variable), the researcher can vary the lighting and record how the reading speed changes at each level.

• To study if reading ability (independent variable) affects first-year college performance (dependent variable), the researcher cannot actually manipulate either variable. Instead, the study must record both variables for the incoming freshman class and form groups based on the student's reading ability.

In both studies, there are multiple variables that must be allowed for in the data analysis. It is not sufficient to just do a t-test of light level versus reading speed or of reading ability versus first-year performance. There are hosts of variables that interact with the independent variable, but few that can be controlled. The first example needs to consider and analyze for effects of age, reading ability, and visual acuity. The second example needs to consider and analyze for effects of age, high school GPA, SAT scores, financial aid level, work schedule, etc.

Between Subjects and Within Subjects

Research studies have two ways of comparing the effect of the independent variable on the subjects.

Between subjects   A subject only sees one research variable option.
　　　　　　　　　　A study on a method of improving grammar usage that has one section use the traditional method and one section use a new method. The final writing sample is compared between the sections. This evaluates different subject's work, so it is between subjects because they do not receive both instruction methods. Likewise, for a web usability study where a subject interacts with the new web page or the old web page (but not both).

Within subjects   A subject sees multiple research variable options. Within subjects designs have more statistical power, but also can suffer from ordering effects.
　　　　　　　　　　A study on a method of improving grammar usage gives a test before and after the session. The same people took the before and after tests, so it is within subjects; the study can directly compare how the subject's grammar usage has changed. A web usability study where a subject interacts with both the new web page and the old web page is a within subjects design; the study can compare how the measured variables (task time, errors, etc.) changed between the two sites.

Ordering Effects

Ordering effects are the result of past study trials effecting the results of the current trial. The basic research design should work to minimize ordering effects and often forces the study to a between subject's design.

In a study on the effect of page layout on reading comprehension where the subject read a text and then answered questions, clearly the subject cannot read the same text with different layouts. There would be a strong learning effect on the second text.

The data analysis of within subject designs must include examining for ordering effects to ensure they are not influencing the results.

A study looked two different designs of a web site and had subjects buy a sweater. Potential ordering effects can affect the following factors:

- Time to decide which one to buy. For the first site viewed, the subject has to determine if this is a sweater they want. On the second site, they can quickly pick it since they have already thought through questions about if they like that style sweater.
- Navigation issues. If the two sites have similar navigation, any potential problems will be encountered and learned on the first one. Those problems will not appear during the use of the second site.

Validity and Reliability

A well-designed study produces data that is both valid and reliable. In common English, valid and reliable are synonyms and more or less mean "true." This is not the meaning they have within research. Lack of either of them severely compromises the study's claims.

Reliability   The extent to which a study yields *the same result on repeated trials*. If the study is run 10 times, it should give the same result all 10 times.

Low reliability could come from lack of good data collection (poor recording practices, poorly calibrated equipment, etc.) that result in wide variations for what should be the same value.

The exploratory data analysis phase can often reveal reliability problems because graphed data shows high data scatter.

Validity   The degree to which a study *accurately assesses the concept* that the researcher is attempting to measure.

Low validity would come from not measuring the variables that should be measured for a study. This could be poorly worded or having leading survey questions. For example, a political survey question worded as "How bad of a job do you think Mayor Smith is doing on fixing the poor streets?" is clearly leading to person to giving Mayor Smith a low rating. A survey such as this one might be reliable (the question is consistently answered the same), but not valid.

A study with high reliability does not necessarily imply a high validity and vice versa. If the wrong variables were measured, the reliable could be high, but what was measured was not what the researcher though was being studied. Likewise, a study with high validity may not have a high reliability if the data collection methods are poor. A study with poor validity is pretty much unsalvageable without a redesign, but a study with poor reliability may just need better data collection procedures.

Variable Types

Quantitative data is almost always associated with a scale measure.

Variables come in two different types: continuous and discrete. The types of data analysis and the statistical tests that are valid for each vary, based on the variable type.

Continuous   Continuous variables are measured along a continuous scale that allows for infinitely fine subdivision. This means they can be divided into fractions, such as temperature, task time, or people's height. With continuous variables, two items can be measured and differences directly calculated. Subject A took 3.43 s longer than subject B to complete a task.

Discrete   Discrete variables are measured across a set of fixed values, such as age in years (not microseconds), year in school (freshman, sophomore, junior, senior), or gender of the subjects. The arbitrary scales used in surveys are discrete variables, such as scoring your level of happiness on a scale of 1–5. Most discrete variables do not support simple math calculation; the level of happiness of a group is not the average of all the 1–5 answers because the differences between values of 1–2 are different from 2 to 3 and 3 to 4. Improperly performing math on these types of values is common.

Type of Data

Quantitative studies by definition deal with numbers. They collect some type of numerical data, perform an analysis on those numbers, and then draw conclusions. However, the numbers that make up the data set are not all the same type. Depending how the data was collected and organized, the numbers fall into one of four groups. The proper analysis depends on the data type; most statistical tests are only appropriate for some data types. This section defines the four types.

Nominal Data

Nominal data does not have a natural order for the categories. Examples might be employee status, gender, race, religion, or sport. Nominal items may have numbers assigned to them, but these numbers are used to simplify capture and referencing of the data.

An employee's department (accounts payable, shipping, marketing, etc.) is nominal data. Numbers could be assigned, but there is no inherent order. It makes no difference in the data analysis if accounts payable $= 1$ or shipping $= 1$. Thus, no arithmetic can be done on those numbers.

Ordinal Data

Ordinal data are data sets where the numbers are in order, but the distances between numbers are unknown. Ordinal data is a discrete variable that has a relative order based its position on a scale. Categorical variables that judge size (small, medium, large) are ordinal date. Attitudes scales (strongly disagree, disagree, neutral, agree, strongly agree) are also ordinal data. People may be grouped by some increasing skill ability (such as Microsoft Word proficiency) into groups A, B, and C.

With ordinal data, the distance between the categories cannot be measured. A common ordinal data is the Likert scale, where $1 =$ strongly disagree, $2 =$ disagree, $3 =$ neutral, $4 =$ agree, and $5 =$ strongly agree. Although these numbers are in order, the difference between strongly agree and agree ($5-4$) is not necessarily the same as between disagree and strongly disagree ($2-1$).

Because ordinal data lacks defined intervals, it does not support arithmetic—it shows sequence only. Thus, calculating averages of single survey questions that used a Likert scale is not valid—but often done and reported. A result of "an average agreement value is 3.6" is using the data in a manner that it does not support.

Interval Data

Interval data is measured with continuous variables using a scale in which each position is equidistant from one another, but does not have a zero that has meaning. Unlike ordinal data, this allows for the distance between two pairs to be measured and compared. Celsius and Fahrenheit temperature scales are interval; if two items are differing by 20° (10–30 and 140–160), then they have the same temperature difference.

Although interval data supports addition and subtraction, it does not support multiplication and division. A temperature of 50 °C is not "half as hot" as a temperature of 100 and a change from 90° to 100° is not a 10% increase in temperature.

Ratio Data

Ratio data is measured with continuous variables with a zero that has meaning. The numbers taught in math classes are ratio numbers. Data collection of ratio data requires that the scale contain values that are equal distance from each other and has a meaningful zero value—unlike interval data, which only requires equal distance.

Observations that can be counted (tasks completed, errors, or height) are typically ratio data. Completing four tasks is twice as high as completing two tasks. With ratio data, numbers can be compared as multiples of one another. Ratio data can be multiplied and divided because not only is the difference between 1 and 2 the same as between 3 and 4, but also because 4 is twice as much as 2. Thus, one person can be twice as tall as another person.

The Kelvin temperature scale is a ratio scale because on the Kelvin scale zero indicates absolute zero in temperature, the complete absence of heat. So, 200°K is twice as hot as 100°K.

Independent Measures and Repeated Measures

A study design can collect data from the subjects in two ways: independent measures or repeated measures. The statistical tests used to analyze independent measures or repeated measures are different for many of the tests. Repeated measures are more powerful but can also exert other confounds such as order effects.

Independent measures	Independent measures research design is when the subjects are divided into separate groups. One group is typically the control and the other is a treatment.
	If the study was looking at the effects of page layout on reading comprehension, with an independent measures design one group would read layout A and another group would read layout B. Since each group only see one layout, they can

contain the same information, eliminating effects of different writing complexity. If a study looked at gender differences in reading speed, the design would be an independent measures design, because it is looking at the difference between male/female performances, so there are two distinct groups.

Most social interaction studies or usability tests must be independent measures since you cannot ask a person to interact with the same information twice without having to deal with order or learning effects.

Repeated measures

Repeated measures research design is when the subjects do the conditions multiple times. Standardized tests can be considered a form of repeated measures testing. The reading comprehension part has you read and respond to several different texts, with the results being an overall evaluation and not score for any individual text. There is a repeated measurement of your reading comprehension.

For example, a study on reaction time may have a subject push a green button whenever there is a stimulus (light flashes, sound is heard, etc.). This can be performed many times for each subject.

If the study was looking at the effects of layout on reading comprehension, in a repeated measures design everyone would read and be tested on multiple sets of texts with different layouts. Or, in a usability test of how well people understand a display, they will answer several (repeated) questions using the display for different situations. Clearly, to minimize order effects, each layout must have different content, and each question needs to be similar in content, but not identical.

Before and after situations are also repeated measures. For example, the subjects take a test on design knowledge before and after a training session. Here, the before test serves as the control and the after test looks to see if the training had an effect. To be repeated measures, the same subjects must take both tests (this is also a within subjects design).

Table A.1 Study design containing both independent and repeated measures.

	Teaching method A		Teaching method B	
	Pretest	Posttest	Pretest	Posttest
Female				
Male				

Each cell in the table must have enough subjects to give the statistical tests adequate power.

Some study designs, especially those with a factorial design contain a combination of independent and repeated measures. In Table A.1, a full data analyses would include both independent and repeated measures analysis. Assume that different people were assigned to teaching method A and B. For example (not all combinations that could be examined are listed):

Repeated measures
- Female teaching method A pretest versus posttest
- Male teaching method A pretest versus posttest

Independent measures
- Female versus male teaching method A posttest
- Female teaching method A posttest versus female teaching method B posttest

Each of the cells in the table need to have enough subjects—as determined by a power test—to give meaningful results. If the teaching method A only had three females, then gender-based analysis would not yield reliable results because of too small of a number.

Variation in Data Collection

Errors are inherent in any data collection procedure. No study can collect data without introducing some variance into the data. In classical measurement theory, any measurement consists of the "true" value and an "error" value. The amount of variation in the recorded values (its standard deviation) reflects a variation in both of these values. Clearly, a good study design strives to minimize the error component. In physics studies that attempt to measure values of fundamental constants, the experimenters work to develop new methods or refine existing measurement devices, so they are more accurate (i.e., smaller error component).

Some error occurs because the equipment is not perfect. A machine gives a reading of $10 +/- 0.2$ units. The recorded value is 10.0, but the actual value may be between 9.8 and 10.2. Better equipment may reduce that error, but it cannot completely eliminate it.

In social science studies, variations between subjects introduce data variation. Factors such as lack of sleep or motivation can cause responses to vary. Or perhaps while collecting data on some subjects, a construction crew was making loud pounding noises that were highly distracting. Researchers work to minimize these errors, but some level of them must be accepted and accounted for in the analysis.

Some errors occur because of researcher errors. Using poorly worded survey questions or questions that do not address what the researcher wants to study introduce errors. Also, improperly recording data (entering 9.24 instead of 9.42) would be an error. These errors can be corrected and potentially eliminated.

Variation in answers

The student course evaluation survey at a university was a list of yes/no questions that were constructed so that the entire class should respond with a yes or a no answer. For example, "the instructor cancelled class more than three times." The instructor either did or did not cancel class this many times, so all the students taking the survey should give the same answer.

Interestingly, there were often a number of students (3–4) who said yes class had been cancelled, while the rest of the class (15+ students) said it had not. Obviously many instructors complained about how impossible this was. From a data analysis view, the answer is not a problem with the survey question, but reveals the inherent variation people give in responding to survey questions. In this example, the discrepancy is clear, but, unfortunately, although this happens in most surveys, it is not easy to spot. Using a one-to-five Likert scale masks the effects of a subject answering all 1 or sequencing through 1–5.

Types of Error

Any statistical analysis risks finding or not finding significance when the reality is the opposite. It may have a false positive—finding significance when it does not exist—or a false negative—finding no significance when it does exist. Typically, these occur because of excess of variation in the data. Table A.2 shows the possible results.

Most of the research methods books address these two types of errors and possible ways of minimizing their effect on the study.

Table A.2 Possible false finding.

	Reality—true	Reality—false
Result: Accept	Correct decision	False positive type II error
Result: Reject	False negative type I error	Correct decision

These differences fall into two categories that are called type 1 and type 2 errors.

Type 1 error	False positive. Where the study rejects the null hypothesis, when reality supports the null hypothesis. In other words, the study found statistical significance when none existed.
	Examples of type 1 errors:
	A medical study looks at new drug for treating a disease. If a type 1 error occurs in the study, it means that the study will say the drug is beneficial for disease treatment when it is not.
	A computer interface is tested to if new design makes it faster. If a type 1 error occurs in the study, it means that the study will say the new design is faster than the old design when it is not.
	A teaching study looks at improving math skills. If a type 1 error occurs in the study, it means that the study will say the new method improves math skills better than current methods when it does not.
	In law terms, convicting the innocent man.
Type 2 error	False negative. Where the study accepts the null hypothesis, when in reality the null hypothesis is false. In other words, the study did not find statistical significance when it should have.
	Examples of type 2 errors:
	A medical study looks at new drug for treating a disease. If a type 2 error occurs in the study, it means that the study will say the drug is not beneficial for disease treatment when it is actually beneficial.
	A computer interface is tested to determine if new design makes it faster. If a type 2 error occurs in the study, it means that the study will say the new design is not faster than the old design when it is really faster.
	A teaching study looks at improving math skills. If a type 2 error occurs in the study, it means that the study will say

> the new method does not improves math skills better than current methods when it actually does.
>
> In law terms, finding a guilty man innocent.

Bias

A bias is a systematic deviation from an expected path (Tversky and Kahneman, 1974). Any set of information interpretations or decisions shows a random spread around the real value; the purpose of a statistical-based analysis is to allow for that random spread. But a bias systematically shifts that spread either higher or lower, so everyone in a group is consistently above or below the average by a set value. Gender studies have found the people reading identical resumes differing only the person's name (John or Jennifer) will score them differently based on perceived gender (Moss-Racusin et al., 2012). The bias has shifted the score for women down.

Information biases are fundamental ways in which people's judgment is consistently affected to produce a nonoptimal decision. In general, it is very difficult for people to adjust for most biases since biases are deeply rooted in the cognitive processes of information evaluation and decision-making. The study of biases is an active area of psychology research that has identified a wide range of specific bias types.

This section describes several different types of bias, but they all result in a tendency to skew the fundamental design of a study or the interpretation of the data. In many cases, a bias can be considered as resulting from a "one-sided perception" by the researcher. It can also be from a one-sided perception by a subject but most studies where that is a major factor would be considering that perception as a study variable, such as looking a people's views of political issues.

A big issue is that a researcher's bias directly influences a study's design and, consequently, can affect the findings. The bias can show itself in many areas of a study, for example, survey question wording, sample collection, what observations should be recorded, or what observations do get recorded.

There are two main types of bias:

Explicit bias	A bias arising for the reader's personal agenda. The researcher assigns values and interprets information to fit a predetermined mindset. For example, a study designed to show a new method is detrimental, to support a political stance, or refute/confirm a previous study with questionable results.
Inherent bias	The bias arising from basic human cognitive processes. Researchers can try to minimize them in a design, but they still exist. The specific examples listed next are all sources of inherent biases.

Some inherent biases of data collect and analysis are listed here. These are biases that can influence the researcher, there is a much bigger set of biases that can influence the subjects and which a design should allow for as part of the data collection and interpretation methodology.

Availability bias
: Collecting or using information that is easiest to obtain. A study on employees may find it much easier to pull information from the HR database rather than observe or question employees directly. But there is no assurance the HR database is highly correlated with the study's variables.

Confirmation bias
: People tend to only seek information that confirms their beliefs or desired decision. A study design may be focused on collecting data that will confirm a belief held by the researcher. Missing from the study is the data that would disconfirm the belief.

Expectancy bias
: When people look at information, expectancy bias causes them to see what they expect to see (Wickens and Hollands, 2000). Klein's research (1999) found that people tended to interpret a passage in one distinct way and that these interpretations were consistent with their backgrounds. If prior knowledge leads people to expect to find specific information, expectancy bias will often result in their finding that information. People reading a study will see what they want to see. Likewise, researchers analyzing the results of a study will tend to interpret it to see the results they want. This may include explaining away disconfirming results as the results of data collection errors, outliers in the data, etc.

There are also several biases that have been found that are based on how people interpret numbers and statistical information in particular. These fallacies are not only applicable to the readers of a study's report; researchers also routinely fall into these biases as part of the data interpretation.

Base rate fallacy
: Overall statistical probabilities are ignored and specific instances are privileged. If people have recently had or heard about a poor or good event in a similar situation, they tend to assume that the current situation will end with the same result. Some people reading study results have this problem and some researchers will want to discard or not believe

data if it fails to correspond to their base rate expectations.

Small numbers bias | Accepting that a small sample is representative of a population. In user testing, the four people tested are often assumed to be representative of the entire user group. If the first four coin tosses come up heads, a coin would be considered biased toward heads. Small samples have wide variances and low statistical power, which must be considered in data interpretation.

Statistical regression fallacy | Exceptional performance (good or bad) tends to regress toward the mean. Any data collected will fluctuate partly by chance, so when there is an extremely good result or score, the next measurement is likely to be worse.

Residuals

Any discussion of statistics eventually uses the term *residuals*. It is the unexplained error in the study. Or more simply, it can be considered as the difference between a data point's real value and the typical or expected value. In general, the lower the overall residual value, the better the data.

Data point = expected value + residual

In simpler terms, the residual means how closely the actual data point corresponds to what is expected. Rarely data points are exactly the expected value; they all show some amount of scatter. The residual captures that scatter and gives a way to consider it in the data analysis. Most statistics software gives a residual plot as part of calculating the regression. Figure A.1 shows the residual for fitting a line in a linear regression (the case where the x value predicts the y value). A graph of the residuals for a data set shows how well the data fits the model. The goal is for the residuals to be small and patternless with an equal number above and below the line.

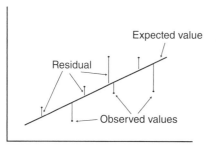

Figure A.1 Residuals for a regression line. The residual is the difference between the expected value and the actual value.

Table A.3 Residuals for student test scores.

Student	Score	Residual
A	80	−5
B	88	3
C	92	7

The residual is the difference between the expected (in this case 85) and the student's actual score.

The statistical independence discussed in the previous section would show an increase in residuals if graphed in the order the subjects performed the task if statistical independence was violated in the study. Notice how, in Figure A.1, the residuals on the left part of the graph are smaller than the residuals on the right part.

The residuals are important in statistical analysis since they are a major component of how the various tests determine statistical significance. In regression fitting, the calculations are working to minimize the residuals as a way of determining the best curve that fits the data.

In testing how well data fits a model, such as during a regression test, the expected value would be the value expected based on the regression equation.

For example, base on various factors (i.e., time studying, previous class grades, and last test grade), three students were expected to score an 85 on a test. Table A.3 shows the residuals for those students based on their actual test scores. Residuals can be positive or negative numbers depending on how they differ from the expected value.

Residual graphs that are skewed show the proposed line is not the best fit. For example, data that fits a power function (curved line) would show a skewed residuals graph (Figure A.2). After running a regression, a researcher needs to examine the residual graph and determine if it looks acceptable.

Figure A.2 Residuals forming a poor fit. Residuals from a linear regression when the data actually need an exponential regression. Notice how the data points track the curved line. For low values, the difference between the linear and exponential function is small, so the residuals are small. For larger values, they increase and shift to all being high.

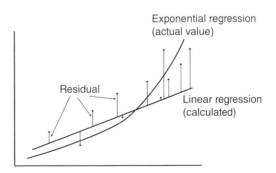

Large residuals are large indicate low quality data and call the reliability of the study into question.

Probability—What 30% Chance Means

Research results tend to be given in probabilities. How people interpret what a probability means and how they respond to different ways of saying the same thing can strongly influence how the research results are perceived.

A statistical analysis results in percentages. Although an obvious sounding statement, many people have a difficult time grasping what those percentages mean. This section helps to clarify what those percentages really mean.

One communication issue with statistics is that most explanations are in terms of events or sampling. An event could mean that there is a 30% chance of a part being damaged or a 74% chance a person likes pizza. Weather forecasts say there is a 30% chance of rain tomorrow, but most people do not understand what that 30% actually means. To fully explain the situation, the rain forecast might explain what area could get rain and how a successful calculation is made. For a drug with a 10% chance of a side effect, the physician needs to explain that 10% of the people taking the drug get a stomach pain as a side effect, and not that once for every 10 pills will a person feel a stomach pain.

The standard for a *p*-value is 0.05, which can be interpreted as a 95% chance the result was not from chance. Or a study might find that 36% of the time people will follow path A. When writing up a research study, it is important to properly interpret the percentages. It is easy to report them in the results section, but it is the discussion section of a paper that has to interpret the numbers and it is there that many people have difficulty in both writing and reading. Too many researchers and readers reveal they lack a clear under-standing of what 36% chance means.

One factor in misunderstanding risks is that the 10% chance of an event typically gets presented without enough context for the person to place it into their situation. People try to use common sense when considering statistics, but they often lack enough experience to have developed the proper common sense. Consider these different examples.

- A person is in a club with 40 members that has a big birthday party for each member. With 40 people, what is the chance of two members having the same birthday? The actual chance is more than 90% that two will share a birthday. [Says the author who graduated in a high school class of 43 and who did share birthday with someone.] The misinterpretation is that many people think of it in terms of sharing their birthday: any person they meet has a 1/365 chance of that. The real question is for any pair of people in the club.

What the 90% really means is that if we looked at lots of groups of 40 people, 90% of them would contain at least one pair who has the same birthday. On the other hand, if picking any specific group of 40 people, you cannot say if there is a birthday pair.

- Most people regard an average as the typical value. For example, when asked how they could use the average temperature for a city, a common response was that it can tell you want to wear during a visit (Garfield and Ahlgren, 1988). In reality, knowing the average yearly temperature of Chicago would be useless in deciding what type of clothes to bring since January and July are very different temperature-wise. But knowing the average monthly temperature in March would be more helpful for a March visit, but the standard deviation is still needed to understand the potential temperature variation.

A "30–50% chance" can be interpreted different ways

A psychiatrist who prescribed Prozac to depressed patients would inform them that they had a 30–50% chance of developing a sexual problem, such as impotence or loss of sexual interest. On hearing this, many patients became concerned and anxious. Eventually, the psychiatrist changed what he said to tell patients that out of every 10 people to whom he prescribes Prozac, three to five experienced sexual problems. This way of communicating the risk of side effects resulted in his patients being more at ease (Gigerenzer et al., 2005).

The basic problem was how his patients understood what "a 30–50% chance of developing a sexual problem" means. It turned out that many patients thought that a side effect would occur in 30–50% of their sexual encounters. The psychiatrist meant 30–50% of the patients would experience a side effect, which also means that 50–70% would not.

Any discussion of percentages must be clear about the baseline of the measurement. In this example, does the percentage refer to a people (patients who take Prozac), to events (a given person's sexual encounters), or to some other factor?

When writing content, probabilities can be described as either a percentage (20%) or a frequency (2 out of 10). Although these numbers are identical mathematically, people tend to respond to them differently. Results written using these two different formats (30% will versus 70% will not), although saying the same thing, are interpreted differently because of framing effects (Levin and Gaeth, 1988; Tversky and Kahneman, 1981). Slovic et al. (2000) asked social workers about the chance of violent criminals repeating a violent crime within 6 months. They found the following:

- Frequency scales led to a lower estimate that a person would commit a violent crime than percentages.

- Frequency scales led to a higher estimate the person posed a higher risk than percentages.
- Frequency scales led to a person being judged as more dangerous than percentages.

The second and third bullets contradict the first one, although the values given to the social workers making the estimates were the same. Slovic, Monahan, and MacGregor suggest that both formats be used when communicating information. "Of every 100 patients similar to Mr. Jones, 20 are expected to be violent to others. In other words, Mr. Jones is estimated to have a 20% likelihood of violence" (p. 285). When writing up a study's findings, the researchers need to take responsibility for clearly communicating the results and not assume the reader will interpret statistical information in the same way the researchers are interpreting it.

References

Garfield, J. and Ahlgren, A. (1988) Difficulties in learning basic concepts in probability and statistics: implications for research. *Journal for Research in Mathematics Education*, **19**, 46–63.

Gigerenzer, G., Hertwig, R., van den Broek, E., Fasolo, B., and Katsikopoulos, K. (2005) "A 30% chance of rain tomorrow": how does the public understand probabilistic weather forecasts? *Risk Analysis*, **25** (3), 623–629.

Klein, G. (1999) *Sources of Power: How People make Decisions*. Cambridge, MA: MIT.

Levin, I. and Gaeth, G. (1988) How consumers are affected by the framing of attribute information before and after consuming the product. *Journal of Consumer Research*, **15**, 374–378.

Moss-Racusin, C.A., Dovidio, J.F., Brescoll, V.L., Graham, M.J., and Handelsman, J. (2012) Science faculty's subtle gender biases favor male students. *Proceedings of the National Academy of Sciences*, **109** (41), 16474–16479.

Slovic, P., Monahan, J., and MacGregor, D. (2000) Violence risk assessment and risk communication: the effects of using actual cases, providing instruction and employing probability versus frequency formats. *Law and Human Behavior*, **24**, 271–296.

Tversky, A. and Kahneman, D. (1974) Judgment under uncertainty: heuristics and biases. *Science*, **185**, 1124–1130.

Tversky, A. and Kahneman, D. (1981) The framing of decisions and the psychology of choice. *Science*, **211**, 453–458.

Wickens, C. and Hollands, J. (2000) *Engineering Psychology and Human Performance*. Upper Saddle River, NJ: Prentice Hall.

Index

Introduction to Quantitative Data Analysis in the Behavioral and Social Sciences,
First Edition. Michael J. Albers.
© 2017 John Wiley & Sons, Inc. Published 2017 by John Wiley & Sons, Inc.
Companion website: www.wiley.com/go/albers/quantitativedataanalysis